U0156301

茶和世界
共品共享

TEA AND WORLD
SHARING AND ENJOYING

农业农村部农业贸易促进中心　组编

Edited by Agricultural Trade Promotion Center
Ministry of Agriculture and Rural Affairs of the People's Republic of China

中国农业出版社
China Agriculture Press
北京
Beijing

图书在版编目（CIP）数据

茶和世界　共品共享：汉英对照 / 农业农村部农业贸易促进中心组编 . —北京：中国农业出版社，2023.4

ISBN 978-7-109-30047-7

Ⅰ . ①茶… Ⅱ . ①农… Ⅲ . ①茶文化－中国－汉、英 Ⅳ . ① TS971.21

中国版本图书馆CIP数据核字(2022)第175287号

茶和世界 共品共享

CHA HE SHIJIE GONGPIN GONGXIANG

中国农业出版社出版

地　　址：北京市朝阳区麦子店街18号楼

邮　　编：100125

策　　划：刘爱芳

责任编辑：李　梅

装帧设计：张今亮　任晓宇　　责任校对：吴丽婷

印　　刷：北京中科印刷有限公司

版　　次：2023年4月第1版

印　　次：2023年4月北京第1次印刷

发　　行：新华书店北京发行所

开　　本：787mm×1092mm 1/16

印　　张：12.5

字　　数：650千字

定　　价：198.00元

谨以此书
献给全球爱茶人

To Each and
Every Tea Enthusiast Worldwide

Special thanks to ROYAL Teahouse Group

序言一

“茶和世界，共品共享”，这是全球茶人的共同愿景。中国是世界茶树原产地，是最早发现和利用茶的国家。目前，全世界有60多个国家和地区种植茶树，茶叶产量近600万吨；茶叶消费遍及160多个国家和地区，饮茶人口超过20亿，国际茶叶贸易量接近200万吨。茶叶是农产品贸易中最为活跃的产品之一，茶产业已经成为很多国家特别是发展中国家的农业支柱产业和农民收入的重要来源。

茶是世界公认的天然健康饮料，茶与健康一直是国际茶叶科学研究焦点。至今，国际上已从体外活性试验、动物实验和人体临床试验三个层面，从细胞生物学和分子生物学水平上探究了茶叶主要功能成分的生物活性及其作用机制，揭示了儿茶素、茶黄素、茶氨酸、茶多糖、咖啡碱等功能成分及不同茶类在延缓衰老、调节糖脂代谢、减肥、调节肠道菌群、调节免疫、抗肿瘤、抗抑郁、抗炎症、抗病毒、抑菌、壮骨骼等方面的作用与机制。不断涌现的研究成果极大地丰富了茶的健康属性，并为茶的健康价值驱动消费提供了可靠的科学依据。

2019年12月，第74届联合国大会通过决议，将每年5月21日确定为“国际茶日”，以赞美茶叶的经济、社会和文化价值，促进全球农业的可持续发展。2020年5月21日，中国国家主席习近平向首个“国际茶日”系列活动致信表示热烈祝贺。习近平指出，茶起源于中国，盛行于世界。联合国设立“国际茶日”，体现了国际社会对茶叶价值的认可与重视，对振兴茶产业、弘扬茶文化很有意义。作为茶叶生产和消费大国，中国愿同各方一道，推动全球茶产业持续健康发展，深化茶文化交融互鉴，让更多的人知茶、爱茶，共品茶香茶韵，共享美好生活。

三年多以来，全球茶人组织开展了丰富多彩、形式多样的教育和宣传活动，为推动茶产业健康发展、深化茶文化交流互鉴不断贡献新的力量。这片来自东方的神奇树叶成就了全世界的举杯共饮。

正是应时代之需，农业农村部农业贸易促进中心组织团队精心编写了本书，具体内容分为五个部分：

第一部分：茶的传奇，一片神奇的东方树叶。通过神农尝百草发现茶，以及历代茶人茶事，以点带面，在读者面前展开了一幅瑰丽绚烂的茶文化历史画卷。

第二部分：茶的经典，伟大的《茶经》。用浅显易懂的文字，描述了世界第一部茶书《茶经》的主要内容，让这部古老的著作持续焕发时代光芒。

第三部分：茶的品赏，天南海北一壶春。全面介绍了各种茶类、饮茶方式、品茶要点和饮茶场景，充分展现了茶的多样和共融，诠释了一杯茶何以让世界更加包容、和谐和美好。

第四部分：茶的传播，和合世界香飘远。展现了茶在全世界落地生根并融入人们的生活，形成缤纷多彩的茶饮习俗和茶文化，带给人们健康、财富与和谐美好。

第五部分：茶的节日，全球茶人的“国际茶日”。记录了自2019年第74届联合国大会将每年5月21日确定为“国际茶日”以来，世界各地举办的“国际茶日”庆祝活动，茶的馨香已弥漫天南海北，全球爱茶人欢庆自己的节日。

一片神奇的树叶，为世人带来了健康，带动了经济，更促进了人类文明与进步，成为构建世界和平的桥梁与纽带。

通读全书，在古今中外茶的世界中徜徉，既能感受到悠远的茶文化历史，又能领略多姿多彩的世界茶俗，还能体会到茶在全世界呈现出的强大生命力，以及茶所肩负的让人类生活更美好的时代使命。本书的出版，一定能让更多的人知茶、爱茶，从而实现“共品茶香茶韵，共享美好生活”的美好愿景，是献给“国际茶日”的最好礼物！

刘仲华

2023年立春

刘仲华，男，1965年出生，中国共产党第二十届中央委员会候补委员，中国工程院院士、湖南农业大学教授。主要从事茶叶深加工与功能成分利用、茶叶加工理论与技术、饮茶与健康等方向的教研工作。

Foreword I

"Tea and World, Sharing and Enjoying" is the common vision of tea lovers around the world. Camellia sinensis (L.), or commonly called tea plants, originated from China. The country is also the earliest to unfold the mystery of tea. Today, tea is a crop for more than 60 countries and regions with a total output of nearly six million tons annually. Now we have a tea-drinking population of two billion from over 160 countries and regions and a global tea trade volume close to two million tons every year. Tea is critical in the world's farming trade, and it is also a pillar of the agriculture and a significant cash crop in many countries, especially in the developing world.

Tea is a well-recognized green and healthy beverage. Ties between tea and health have always been a big subject for tea scientists. For the biological activity and mechanisms of the main functional components of tea, cell and molecular biologists have done great jobs, from the perspectives of in vitro activity, animal experiment and human clinics, to unveil how catechins, theaflavins, theanine, tea polysaccharides, theine and other components in various tea types work to help delay the aging process, lose weight, strengthen bones, suppress tumors, depression or inflammation, kill viruses and bacteria, and regulate immunity, intestinal flora and glycolipid metabolism. The prolific research findings have expanded the list of tea's health effects and can help, with such a strong evidence for its health value, back up and give a push to the global tea consumption.

The 74th session of the United Nations General Assembly in Dec. 2019 adopted a resolution to name May 21 as "International Tea Day" to celebrate the economic, social and cultural value of tea and promote the sustainable development of the global agriculture. On May 21, 2020, Chinese President Xi Jinping sent a letter to extend his warm congratulations to the celebrations of the first International Tea Day. Xi noted in his letter that "tea originated in China and became popular in the world. The United Nation's decision to set up the International Tea Day showed the international community's recognition of and emphasis on the value of tea, and the move is of great significance to revitalizing the tea industry and carrying forward tea culture." The president also stressed that "China, a major producer and consumer of tea, is willing to work with all sides to nurture the sustained and healthy development of the global tea industry, deepen cultural exchanges on tea, and allow more people to relish lives accompanied by tea."

The global tea community have, over the past three years, carried out a variety of education and publicity sessions for the future development of tea industry and mutual understanding of tea cultures. The magical leaf from the East is in the spotlight.

To respond to the call of times, the Agricultural Trade Promotion Center of the Ministry of Agriculture and Rural Affairs of the People's Republic of China prepared this book with five chapters as follows.

Chapter One is "The Saga of Tea: A Magic Leaf from the East". The splendid tea culture of China came onto the stage ever since the plant was discovered by Shennong 5,000 years ago. Readers will get a grip of the entire tea history with significant people and events from various periods.

Chapter Two, "The Literary Masterwork of Tea: The First, and the Greatest", explains to readers in simple terms the ancient tea knowledge and presents a whole picture of the very first tea monograph of the world.

Chapter Three is "The Knowledge of Tea: A Beverage for the Global Village". This chapter will walk you through the types, scenarios and drinking methods of tea, plus useful tips, explaining why a cup of tea makes the world more inclusive, harmonious and beautiful.

Chapter Four is "The Spread of Tea: A Rich Smell to Please the Entire World".

Tea has taken root all over the world, integrated into people's lives, formed colorful tea drinking customs and tea culture, and brought people health, wealth and happiness.

Chapter Five is “The Festival of Tea: A Special Day for the Global Tea Community”. It records the celebrations around the world since the 74th General Assembly of the United Nations in 2019 designated May 21 as “International Tea Day” every year.

The magic leaf brings health to the world. It also drives the global economy and human civilizations, and is often regarded as a bridge to world peace.

You will see from this book the long history of tea culture, diverse tea customs, vitality of tea industry in the world and people's aspiration for a better life. I hope readers will learn more about tea, fall in love with tea, and scent their days with its sweet aroma. The book is undoubtedly the best gift for International Tea Day!

Liu Zhonghua

The Beginning of Spring (the first Solar Term of a year), 2023

Liu Zhonghua (1965-), alternate member of the 20th CPC Central Committee, Member of Chinese Academy of Engineering, is a professor of Hunan Agricultural University. He’s been engaged in the teaching and research on tea deep processing and functional components utilization, tea processing theory and technology, tea drinking and health, etc.

序言二

今天您喝茶了吗？

茶，自中国西南古巴蜀原始森林诞生，经七八千万年的岁月历练，被距今五千年左右的炎帝部族领袖神农首次发现，开始了人类与茶休戚与共的历史。公元 8 世纪的唐朝，陆羽以一部七千余字的《茶经》确立茶的文化地位，开门见山曰：茶者，南方之嘉木也。此处之茶，既可理解为生态环境下生长在南中国的木本植物，亦可诠释为人文语境中君子风范般的嘉物。茶圣的提纲挈领，将茶物质与精神水乳相融的复合形态一语点透，真可谓一叶双菩提。

本书正是基于这样的人文立场，以一片鲜茶作为钥匙，开启地球上这扇绿色的山野之门，就此引导我们的感官如鲜花一样盛开，欣赏和体验绿水青山的满坡馨香。

茶是生活的艺术，它由人类的日常体验起步构成，所以它貌似浅显，实则博大精深。仅从“茶”字的解读，我们对这种源于生活的艺术魅力就可窥一斑。

从“茶”字的汉语发音中，我们通过一个美丽的故事，可以看到茶与语言的关系。传说上古时期的神农长着一个透明的肚子，肠胃里一旦出现了不干净的毒素，人们就能看得清清楚楚。这时我们先民想出办法，就是将一种绿色的长青植物叶子吃进肚子，让它随着肠胃的蠕动不停地擦洗着肚子，直到毒素全部被擦洗干净。为了纪念这种叶子的功效，人们用擦洗的“擦”字的发音来命名这种树叶，从此这种叶子就被叫作“茶”。

从“茶”字的笔画解读中，我们更可以看到中国文化对养生、长寿的高度重视。有一种高寿被称为“茶寿”，就是因为拆开这个“茶”字，可以发现草字头像是二十，“木”字像是八十，而“人”字像是个八，三者相加是108，所以，茶寿便是“108”岁。

再如从“茶”字的字形解读，它最生动直接地体现了中国古老哲学中“天人合一”的生命观、宇宙观和价值观。“茶”由三个汉字构成，分别是“草”“木”“人”，合在一起，“茶”字即草木当中的一个人，何等生动与浪漫！

茶文化的核心构建于中国茶文化背景下的四个文化层面。

一是茶习俗。它建立在人类日常生活基础的行为文化层面，其内容为人际交往中约定俗成的茶文化习俗，柴米油盐酱醋茶可作为关键词，它包括了各国各地各民族之间的礼俗民俗风俗等形态，我们往往可以用民俗学、人类学、历史学、考古学、传统中医学等学科的研究方法去观察其行为模式。

二是茶制度。它建立在人类社会生活的制度文化层面，内容囊括人类在社会实践中组建的各种社会行为规范，涉及茶生产和流通过程中所形成的生产制度和经济制度，包括茶政、茶榷、纳贡、赋税、茶马交易，以及现代茶业经济和贸易制度等。我们多从历史学、政治学、经济学、生态学、管理学等学科角度去着手研究这一层面。

三是茶美学。它建立在人类精神生活的审美文化层面，我们往往以琴棋书画诗酒茶来概括其内容，茶美学着眼以茶的品饮艺术，强调茶的审美实践与品味，内容涵盖茶文学、茶艺术，茶空间、茶器物，茶品牌、茶技艺，茶非遗，尤其是在此审美历程中诞生的中国茶语，即关于中国茶的话语体系。它包括茶自身的物性表达，与茶相关的人类记忆解读，涉茶物质形态的文化诠释等。我们多从美学、文学、艺术学、传播学等领域出发，去学习和实践这一文化层面的内容。

四是茶意识。它建立在人类精神活动孕育出的思维方式文化层面，是茶文化的核心和文化尖端，是人类对茶形而上的价值观念的终极思考。其内容包括了茶与人类信仰、茶哲学观、茶科学观、茶历史、茶教育传习等。我们多从哲学、宗教学、历史学、教育学、自然科学史等诸多学科视角出发去研究这一层面。

以上四大茶文化层面，构成了茶文化的金字塔模式，即以茶习俗为文化地基、以茶制度为文化框架、以茶美学为文化呈现、以茶意识为文化灵魂的茶文化知识体系。

这部茶文化读物正是从这四大茶文化层面入手，既非高头讲章，更无庙堂之气，它一段段地把众多故事和知识科普给想要了解茶的人们，而这正是进入茶领域的最佳途径。这也是一部关于“爱与和谐”的教科书。茶从大地生长，博大精深，没有爱之心灵，是无法真正走近茶的。故本书强调阅读者自“爱茶”伊始，从体悟与感知入门，进入茶与人类互动的情理教化通道。茶是中国人的良心，它崇尚善良、自尊、独立和宽容，它内蕴神秘、自然、灵动和智慧；茶里凝固着中国人的基本人性，是来自中国幽深历史中的中庸和含蓄，温绵和柔韧，代表着社会的正面价值取向。它把中华文明集于一叶，溶于一杯，青枝绿叶，行遍全球，奉献人类，故一百多个国家的数十亿人饮茶，他们是和谐哲学的实践者。

“休对故人思故国，且将新火试新茶，诗酒趁年华。”每年 5 月 21 日，是联合国命名的“国际茶日”。这小小的一片叶子，为什么会有这么大的魅力，吸引全世界的目光呢？因为它是真善美的象征，是中华民族文化基因和密码的承载体，是全世界爱好和平的人们的钟爱。

请从阅读这部书开始吧，把世界铺成一张茶席，请全人类在此共饮。

王旭烽

2023年立春

王旭烽，女，1965年出生，国家一级作家，代表作《茶人三部曲》，获第五届茅盾文学奖。

Foreword II

Tea, please.

Tea, or originally a plant called Camellia sinensis from the distant past, had remained unnamed in the primitive forests in southwest China for tens of millions of years. It was first discovered by Shennong, the leader of the late Emperor Yandi's tribe about 5,000 years ago, when the legend started. China's tea culture took shape first in *The Classic of Tea,* or *Cha-Jing*, a masterpiece Lu Yu wrote in the 8th century. The book begins with Lu's comment that "tea is a magnificent tree from the South." In his eyes, the beverage was more than a shrub species deep in southern mountains. It was an epitome of the highest moral standards for the noble class in China. Lu Yu is revered as the "Sage of Tea". He found the value of tea and further defined it as the avatar of "men of honor", or in short, a heavenly existence of spiritual and cultural significance.

The book can thus walk you through the romance. The magic leaf from the East is a key to find out what's behind the doors. It provides guidelines by which you would pick up something: the plant, the beverage, and the culture.

Tea is an art in life. It simply stemmed from daily use, but as being endued with more values and humanistic anticipations over time, it is now a mystic name. Let's see how we interpret it.

A story about why it is pronounced "cha" in Chinese. Legend has it that Shennong who discovered tea first had a transparent belly. It helped people check if any toxins appeared in his intestines. They urged him to swallow the leaves of an evergreen grove and let them scrub his body with the peristalsis of intestines until all toxins were cleaned off. To remember the leaves' magic, people called it "ca" as pronounced for the meaning "wipe off". It sounds similar to "cha" and was later in wide use.

For a Chinese character, its strokes matter for something. The character "cha" (茶) implies a culture that attaches great importance to health and longevity. In China, 108 years old is called "the age of cha". Why? Split the character, and you will find it composed of three parts. One stands for the age of 20 years old, the second for 80, and the third for 18. Altogether, they represent an age of 108.

It can also be explained from the character's structure, which tactfully reflects the Chinese ancestral philosophy of life, universe and value. "Cha" (茶) is actually composed of three Chinese characters, from head to toe, "grass" (艹), "people" (人) and "wood" (木). See? "People" is sandwiched between the grass and woods around. What a life!

Well, let's know more about the core of China's tea culture in four respects.

First, customs. The variegated tea customs are all based on people's behaviors in life. They took shape with the growing interpersonal exchanges and talks. Everyone knows the "seven things for a good life": firewood, rice, oil, salt, soy sauce, vinegar and tea, which do represent an age-old wisdom in China. Tea aside, a custom may incorporate etiquette and folklore. It's about a range of research fields, like anthropology, history, archaeology, traditional Chinese medicine and the like.

Second, system. Tea customs primarily came from cultural institutions, which refer to behavioral norms and economic systems in the process of tea production and circulation, such as tea administration, tea tax, tribute tea, tea-horse trade in the distant past and modern tea economic and trading system. It's about history, politics, economics, ecology and management.

Third, aesthetics. There must be a tie between tea customs and arts. In ancient China, the noble class couldn't live without a few "life funs", like music, chess, calligraphy, painting, poetry, wine and tea. The art of tea stresses that, instead of "drinking", we have to "sip" for the true flavor of tea. This is a universal topic, covering literature, art, space, utensils, brands, techniques and intangible cultural heritage. We have an established set of terminology about tea and tea culture, involving tea's symbolic values, the tea history with mankind's and the cultural interpretations of teas in varied forms. It's about studies on aesthetics, literature, art and communication.

Fourth, consciousness. It is the core of tea culture, standing as human's mental activity of a paramount level, or a metaphysical understanding of tea of all time. The tea consciousness covers in fields like belief, philosophy, outlook of science, history and education, etc. To learn about it, we'll put eyes on philosophy, religion, history, pedagogy, history of natural science and so forth.

There is a pyramid for the four dimensions. It puts tea customs as the base and is framed by tea rules and systems. Besides, we've developed well-accepted aesthetic values and made tea consciousness a backbone of the culture.

The book I preface is the one that explains the four in simple words. It's full of legends, anecdotes and expertise, a good start to visit the tea world. For me, this is also a classic for "love and harmony" education. Tea is a spirit down from the earth. We love the earth, we love the nature, and therefore we love tea. You would find between lines what to do to love it. In fact, tea is the conscience of the Chinese people. We've been longing for kindness, self-respect, independence and tolerance. It's a Chinese wisdom. Tea is also an epitome of the Chinese people's personality that we're self-controlled, moderate and resilient. They are all positive values. We love a state of peace and balance, or "harmony" as mostly said. Now two-thirds of the world's population are tea consumers. They are, in some ways, all practitioners of such a philosophy.

May 21 is the UN's International Tea Day. For such a tender, green leaf, why does it draw worldwide attention? Because it is a symbol of truth and beauty. It carries the genes of Chinese civilization and has the power to decode it, so it becomes a favorite of so many people around the world.

Let's start with this book. The best way to meet people and make them friends? A cup of tea, please.

Wang Xufeng

The Beginning of Spring (the first Solar Term of a year), 2023

Wang Xufeng (1965-) is a renowned writer. She is a Mao Dun Literature Prize winner with her masterpiece *The Trilogy of A Tea Family*.

茶
5.21
International Tea Day

目录　Contents

茶的传奇
一片神奇的东方树叶

茶的经典
伟大的《茶经》

茶的品赏
天南海北一壶春

茶的传播
和合世界香飘远

茶的节日
全球爱茶人的“国际茶日”

Contents

Chapter One

The Saga of Tea
A Magic Leaf from the East

Chapter Two

The Classic of Tea
The First, and the Greatest

Chapter Three

The Knowledge of Tea
A Beverage for the Global Village

Chapter Four

The Spread of Tea
A Rich Smell to Please the Entire World

Chapter Five

The Festival of Tea
A Special Day for the Global Tea Community

Chapter One

—

The Saga of Tea

A Magic Leaf from the East

第一部分

茶的传奇

一片神奇的东方树叶

引文

一片嫩绿的树叶，静静地生长在中国葱郁的深山林莽之中，5000多年前被神农氏偶然发现，自远古食用解毒始，这片茶叶飞出深山，渐至飘入茶碗，“细煎慢品”成清雅甘露，古代文人于蟹眼乳花之间乐此不疲，怀悠思古、升华心灵。这片树叶由身至心，滋养中华儿女几千载。

不仅如此，1000年以前，这一片嫩绿开启了不寻常的全球之旅，从中国出发，历千余年，在朝鲜半岛、日本、荷兰、英国、印度、斯里兰卡、肯尼亚等地落地生根。至今，在全球五大洲，飘落的茶叶次第展开一个个美丽的故事，一次次影响着世界。

一部陆羽的《茶经》彻底开启了世界茶饮文化，并由此展开了一段至今未完的、一片嫩绿树叶的故事。

Introduction

It was not until over 5,000 years ago when Shennong (Divine Farmer, a deity in Chinese mythology) chanced upon a heavenly green, which otherwise should have remained unnamed deep in forests in the southwest of China. He took it to kill toxins at first. As centuries went by, however, the green retired from the thick woods and, by degrees, became an earthly drink named "tea". It was mindfully cooked to pursue an ethereal brew and, as tea knowledge continued to gain ground, was reckoned a blessing to purify one's heart. In short, tea is a natural gift to the Chinese people, both for body and soul.

The magic green was even on a journey worldwide over the past millennium, sinking its roots in the Korean Peninsula, Japan, the Netherlands, the United Kingdom, India, Sri Lanka and so forth. Now a drink all over the continents, it left myriads of legends that have from time to time changed the world.

So broad and profound, tea culture was first unveiled to the world with *The Classic of Tea*, a masterwork of Lu Yu from China, thus giving rise to a legendary story all about this green. Yet, as we believe, the story will never die.

发现茶的最早传说，神农尝百草偶遇茶叶

前 3000 年

茶叶被发现的传说充满了古老东方的浪漫！神农是中国人心中的大英雄，是农神，也是药神。5000 多年前，他攀山登崖，采集食物，遍尝百草，教人农耕与采药治病。相传神农在巴山峡川中发现了“神奇叶子”——茶，用它为自己解毒，滋养了中华儿女。如今，茶香飘全球，20 多亿人每天都在品尝着这种远古的味道。

神农发现茶的传说，体现了中华民族对生命与健康的信仰与追求。神农的传说不仅仅是神话，现代茶树种质资源分布状况的考察以及在云贵、巴蜀地区大量古茶树的发现研究证明，茶树原产于中国，中国最早发现和利用了茶。

3000 BC

The First Encounter in A Chinese Legend

It seems a chance encounter. Shennong, in Chinese folktales, was a hero 5,000 years ago. He was regarded as the god of both farming and medicine.Shennong tried plenty of plants and herbs on a long trip searching foods around, and hence started to teach his tribe farming and herbal medicine. Deep in Bashan Mountains, as legend says, he chanced upon a magic plant. He used it to save himself from poisoning. The plant, which helater called "tea", thus became an indispensable part of life for the Chinese nation and even people all around the world. For the population of two billion today, tea is a heavenly gift from the distant past.

Shennong's deeds actually stand for the Chinese belief in life and health. The story was not all fabled. The distribution of modern tea germplasm resources, coupled with the crowds of aged tea trees found in Yunnan, Guizhou and Sichuan, gives proof that tea originated in China and was first discovered and used by the Chinese people.

茶作为商品开始交易，西汉人将买茶写入“劳动合同”

前59年

公元前59年，汉代著名文学家王褒到朋友家拜访，让友人的童仆便了去买酒，刁蛮的便了抗命，王褒一怒之下要买下便了管束，便了顶撞说：要使唤我的事项都要写在契约上，没写的，我就不做！于是，王褒写下了《僮约》，作为“劳动合同”，事无巨细地罗列了便了需完成的工作及详细的规矩。《僮约》中有“烹茶尽具”“武阳买茶”两句，说明2000多年前“烹煮茶并将用过的茶具清洗干净”为文人日常生活，而且茶叶已经成为商品，可以在市集上自由买卖。这是中国也是全世界最早的关于买卖茶叶的记载。

Tea and China's Earliest "Job Contract"

Wang Bao was a literary celebrity during the Western Han Dynasty (202 BC-AD 8). In 59 BC, Wang visited his friend and asked a boy attendant of his friend's named Bianliao to buy wine. The unruly boy, however, refused his request. In a rage, Wang proposed to buy Bianliao out for tight control. The boy refused to obey, saying "you have to put all my jobs in a writing. I won't do anything else!" Hence came his contract with Bianliao specifying all jobs to do and rules to observe. Wang Bao particularly stated in this contract that the boy "shall cook tea and clean the tea sets" and "shall buy tea in Wuyang county". The former was actually a part of life for privileged scholars 2,000 years ago, yet the latter reveals that, back then, tea was allowed for free trading in fairs. This is the earliest account of tea shopping in China and even the entire globe.

59 BC

茶叶入藏，文成公主将茶叶带上“世界屋脊”

公元641年，唐太宗李世民将文成公主嫁给了吐蕃（今西藏）的赞普松赞干布。公主丰厚的嫁妆中除了奇珍异宝，还有如今藏族人一日难离的茶叶。

藏族人民都喜欢喝茶，尤爱酥油茶。制作时，先将茶砖解散，煮成浓浓的茶汁，再与酥油、奶、盐等混合，倒入专门用于打酥油茶的茶桶中抽打成水乳交融的酥油茶。一碗热腾腾的酥油茶不仅解渴、饱腹、解乏、暖身，更能减少高原生活缺少蔬菜带给藏族人的身体不适，促进消化，“茶是血，茶是命”“宁可三日无粮，不可一日无茶”，这些藏族谚语体现了藏族人将茶视如生命。

641年

A Gift Bestowed on the World's "Third Pole"

641

Princess Wencheng was married by Emperor Taizong to Songtsan Gambo, King of Tibet in AD 641. Priceless treasures aside, tea was also a part of the princess' elaborate dowry. It is now a must in Tibetan life.

Tibetan people love in particular the buttered tea. To make it, first crack a tea brick and cook the fragments for concentrated soup. Mix it with butter, milk and salt, and then filla customized long barrel with the mixture for stirring and pulping. The pulped mixture will be the buttered tea. Buttered tea is high in calories and therefore energizing, and it also helps resisting cold and digesting high-altitude food. It is also considered an elixir to ease the health problems arising from the less intake of vegetables on the plateau. Tibetan proverbs, such as "Tea is blood and life" and "I'd go three days without food rather than a day without tea", also give proof of the role of tea in Tibetan people's mind.

“茶道”，皎然写诗说茶论道

“孰知茶道全尔真，唯有丹丘得如此。”皎然创作的诗歌《饮茶歌诮崔石使君》中，首次出现“茶道”一词。

皎然是唐代的著名诗僧，约生活于720—803年，留有400多首诗歌。他是陆羽的好友，一生淡泊名利、坦荡豁达，不喜迎来送往的世俗应酬，嗜好饮茶。在《饮茶歌诮崔石使君》一诗中，皎然细腻生动地描绘了剡溪名茶的色、香、味，感叹香茶赛过仙饮琼浆，并描述了自己饮茶感受的几重境界：“一饮涤昏寐”，令人神清气爽；“再饮清我神”，如飞雨洗涤尘埃；“三饮便得道”，破除烦恼，思想得到升华。这首诗完美生动地展现了茶饮带给人的物质功能与精神享受。

720—803年

The Ultimate Truth of Tea
A Monk Proposed

720-803

"Anyone could understand the final truth in tea? Well, probably only the immortal Dan-qiu-zi." It was in "Ode to Tea for Governor Cui", a poem of master Jiaoran, that the name "cha-dao" (truth in tea) first appeared.

Jiaoran (circa AD 720-803), a noted monk-poet of the Tang Dynasty, wrote over 400 poems across his lifetime. He was a friend of Lu Yu, the "Sage of Tea". Open-minded and indifferent to fame or wealth, Jiaoran lived a restful life free from mundane social connections and exchanges. The poem was a review of the color, fragrance and taste of the tea from Shanxi, which in his eyes far outweighed any heavenly wine, and an elucidation of his state of mind over a few cups. "The first brew made me feel so refreshed, and another brew cleaned my thoughts like a sudden rain. With the third I seemed to acquire the ultimate truth, that worldly troubles were all manmade and meaningless." This is a masterful description of tea's benefits for both our body and soul.

茶圣著《茶经》，陆羽著成茶学第一书

陆羽生活于733—约804年，是唐代茶学家，被后人尊为“茶圣”“茶神”。陆羽爱茶，常与友人品茶鉴水、谈诗论道。754年，陆羽开始他的“问茶”之路，用脚步丈量各名茶产地，经过20多年的茶事考察、学习和生产实践，陆羽著成世界第一部茶叶专著《茶经》，成为世界茶学的创始者。

《茶经》刻印问世于780年。全书共分三卷十章，共7000余字，系统总结了中国唐代及唐代以前有关茶叶的科学和实践经验，为后世中国和世界茶业发展做出了巨大贡献。据统计，在中国及日本等地，《茶经》有近60种版本流传于世。

733—约804年

The World's First Tea Book

Lu Yu (AD 733-circa 804), a tea connoisseur in the Tang Dynasty, is respected as the "Sage of Tea" or "God of Tea". Instilled with an affection for tea, he oftentimes brewed tea with friends and took delight in such tea parties. Lu Yu's trek in quest of tea understanding started in AD 754. He spent over 20 years traveling all over the tea-growing areas in China, pursuing knowledge with regard to tea farming, production and customs etc. To summarize his learnings, Lu Yu wrote the world's first tea book *The Classic of Tea (Cha-Jing)* and thus became globally the founder of tea culture.

733-804 (c.)

The Classic of Tea first saw the light of day in AD 780. Comprised of some 7,000 characters, the book has ten chapters that span three volumes. It is a comprehensive introduction of tea studies and practices during the author's time and even pre-Tang dynasties, and is also a monumental contribution to the burgeoning tea prosperity in China and the world. The book has as yet had nearly 60 editions in China and other countries, including Japan.

七碗茶歌，诗仙卢仝醉清风

《七碗茶歌》被后世爱茶人奉为经典，是流传最为广泛的茶诗之一。其实，《七碗茶歌》只是《走笔谢孟谏议寄新茶》的中间部分，作者是唐代爱茶成痴的诗人卢仝（约795—835年）。

《七碗茶歌》描述了品饮七碗新茶的美妙意境——第一碗润喉咙；第二碗除烦闷；第三碗让人文思泉涌；第四碗让人将一切不平事都抛到九霄云外，表达了诗人舒畅豁达的洒脱胸怀；第五碗仿若脱去肌骨，超越凡俗；第六碗通仙灵；喝到七碗时，两腋生风，自己仿佛成了神仙，欲乘清风归去，妙不可言。

《七碗茶歌》将茶解渴、润喉的功能上升至洗涤心灵、豁达心胸的精神层面，用优美传神的诗句表达了卢仝对茶的深切热爱和舒畅豁达的洒脱胸怀，被后人广为引用。

约795—835年

From Comfort to Completeness Just over Seven Cups

"*The Seven Cups of Tea*", a classic to tea connoisseurs, is among the best-known tea poems of all time. In fact, it is an excerpt of the poem "On the Tribute Tea from My Friend Meng Afar". The author is Lu Tong (circa AD 795-835), a poet known for being so besotted with tea in the Tang Dynasty.

The poem reproduces his wild feeling sover seven cups of tea. As he described, "the first cup moistens my throat; the second shatters all feelings of solitude; the third purifies my digestion, re-opening the five thousand volumes I've studied and bringing them to mind afresh; the fourth induces perspiration, evaporating all of life's trials and tribulations; with the fifth cup the body sharpens, crisp; the sixth cup is the first step on the road to enlightenment; and the seventh cup sits steaming - it needn't be drunk, as one is lifted to the abode of the immortals." The verses resound with a deep and timeless spirit, celebrating tea's ability to inspire him to soar.

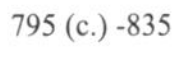

In "*The Seven Cups*", tea's effects were slowly rising from a physical sensation onto eventually the author's communion with his own soul and energy. Lu's beautiful and timeless words, as truly a pledge of his attachment to tea drinking and broad-mindedness in life, are often remembered at tea parties.

以茶礼佛，法门寺地宫出土皇家茶具

中国西安的法门寺出土的文物中有一组唐代宫廷茶具，通过它们，我们得以窥见盛世大唐宫廷茶事的典雅隆重。

地宫出土的账物碑详细记录了大唐君王"以茶礼佛"的规格，出土的数十件珍贵茶器则展示了唐代宫廷茶仪茶道——银丝结条茶笼和鎏金镂孔银茶笼、银质鎏金茶碾和银碢轴、银茶罗、银茶则、银龟盒、鎏金银盐台和素面银盐台、琉璃茶托和茶碗、长柄银匙等，还有极为罕见的秘色瓷茶碗等御用器物。这些器物与陆羽《茶经》中描述的唐代茶事用具基本吻合：茶笼用于盛茶炙茶，碾和碢轴用于碾茶，罗用于筛茶，龟盒、盐台用于贮盐，匙子用于放盐、击拂茶汤。

大唐皇帝以御用茶事器具虔诚礼佛，足见茶事在唐朝人心中的重要地位。

约 874 年

The Unearthed Teaware for Emperors' Ceremony

Alongside scores of cultural relics unearthed at Famen Temple in Xi'an, China's Shanxi province, a set of teaware exclusive to the imperial court of the Tang Dynasty were made public again. They are truly a mirror of the majestic tea ceremony in the prime of Tang China.

A stone tablet unearthed from the temple's underground palace was inscribed with a list of the offerings from the emperors, and dozens of the priceless teaware reproduced the tea rite and ceremony of Tang's imperial court, including the silver strip-knotted tea basket, gilt follow-out tea basket, silver-gilt tea roller set, silver sieve, silver tea-spoon, silver turtle-shaped box, silver-gilt (and plain) spice holder, glazed saucer and bowl, long-handled silver spoon and, most importantly, the rarely-seen Mi'se porcelain tea bowl. Basically, the discovery chimed in with the utensils recounted in Lu Yu's book. In his notes, people of the Tang Dynasty used basket to pack tea, a roller set to grind tea, seize to screen tea, turtle-shaped box and spice holder to contain salt and spices, and spoon to put salt and whisk tea soup.

874 (c.)

The emperors' teaware for Buddhist rite attests to the exceptional part tea played in people's life back then.

天地人和，三才碗蕴含哲学思想

盖碗又名“三才碗”，上面有盖，喻天；中间有碗，喻人；下面有托，喻地。“三才”取天、地、人相互依赖共生，天地人和的喻义。

相传，碗下的托是唐代西川节度使崔宁的女儿发明的，用来防止被茶水烫到和茶杯倾倒。到明代又有人在碗上加盖保温，于是“三才聚首”，盖碗诞生。清代以来，盖碗非常流行，它具有较强的实用性和仪式感，盖子防尘、保温，底托隔温、防烫，还能令茶碗不易倾斜泼洒茶水。喝茶时一手连托端起茶碗，一手打开碗盖，用碗盖轻轻将漂浮的茶叶拂去，俯首啜饮茶水，饮毕盖好碗盖，轻放桌上。

一件茶具的命名体现了中国传统哲学中“天人合一”的思想，反映了中国人器用之道的哲学观。

A Set Comprising the Heaven, the Earth and Mankind

Gaiwan, or "Sancai Bowl", comprises a lid, a bowl and a saucer. The lid represents the heaven, the bowl stands for mankind, and the saucer implies the earth. It suggests the co-existence of the three imageries on the lookout for ultimate harmony.

The saucer was allegedly invented by the daughter of Cui Ning, the Governor of Xichuan in the Tang Dynasty, to prevent the hot brew from burning fingers and the bowl from being upset. It was not until the Ming Dynasty that the lid appeared to keep the brew inside warm, thus finalizing the three-piece set. It became a prevailing device in the Qing Dynasty. Being customarily a pledge of hospitality and gracious manners, gaiwan is no ornament. The lid keeps tea inside dust-free and warm, while the saucer shields our fingers from the heat and holds the bowl well planted. To drink from gaiwan, hold the saucer in the palm with the bowl on it, and gently lift the lid by the knob with fingers of another hand. Carefully incline the lid that can hold back the steeped leaves, take a sip, then cover the bowl with the lid and gently place the set back on the table.

The name "gaiwan" reveals the thought of "the unity of heaven and mankind" in the ancient Chinese philosophy, and mirrors the philosophical outlook of the Chinese on using wares.

精美贡茶，蔡襄改制“小龙团”

贡茶起源于3000多年前，于唐代形成制度。唐大历年间，顾渚建立中国历史上第一座官焙贡茶院，并于770年开始制作贡茶。至宋朝，宋太宗要求贡茶“取象于龙凤”，并在福建建安郡（今建瓯市）设立北苑贡茶院。

蔡襄是宋代著名书法家和文学家，著有《茶录》，一生爱茶。他在任福建路转运使时，将一斤8饼的龙凤团茶改制成一斤20饼的“小龙团”，贡茶品质更佳，备受朝廷推崇，臣子极难获得赐赠，偶尔有幸分得一角团茶的大臣也舍不得饮用，而是孝敬父母或家藏为宝。

宋朝宫廷对茶的珍爱和推动促进了制茶工艺、点茶技艺等的提升和茶具的发展，并带动了全社会的斗茶、品茶热潮。

1046—1048年

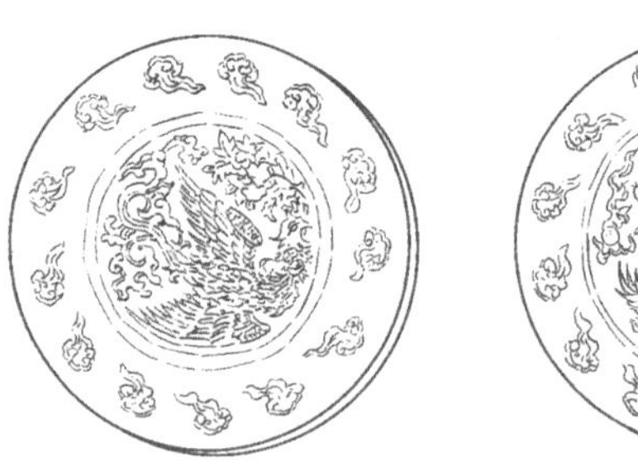

龙凤团茶纹样

The imperial cake tea emblems

The Legend of Tribute Tea and Cai Xiang's Reform

Gongcha, or tribute tea, was the tea of premium quality presented by the local authorities to the imperial court. It first appeared 3,000 years ago, and it was not until the Tang Dynasty that the system began to take shape. China's first official roasted tribute tea workshop was founded by Gu Zhu during the Dali Period (AD 766-779). It started to produce tribute tea in AD 770. The tea was later deliberately inscribed with the auspicious imperial emblems at the request of Emperor Taizong of the Song Dynasty, and it was under his reign that the Beiyuan Tribute Tea Workshop was established in Jian'an(now the city of Jian'ou), Fujian province.

Cai Xiang was an esteemed calligrapher and writer in the Song Dynasty. As the author of "*Cha-Lu* (*The Record of Tea*)", Cai took a great liking to the magic leaves all his life. By then the Transport Commissioner of Fujian, Cai Xiang pioneered the manufacture of a sort of smaller cake tea of superlative quality. Instead of previously eight cakes a pack, Cai filled the pack with twenty cakes. It was called "Xiao-Longtuan", or smaller imperial cake. The new tribute was considered such a rarity that even ministers would have very slim chance to own it. If someone, fortunate enough, was given a fragmented cake, he would rather leave it to parents or store it carefully back home.

1046-1048

The tea processing, techniques and teaware making, in the Song emperors' exemplary support, underwent a noted progress. Alongside the reigning court, there also came a nationwide upsurge of tea contesting and drinking popularity for people from all walks of life.

宋茶点饮，宋人的斗茶雅好

斗茶就是通过“斗色”“斗浮”品鉴茶叶品质，论定胜负。

宋代茶叶仍以饼茶为主流，但茶更精工细制，饮用方法从唐代的煎煮法变化为点茶法。点茶前需先研磨茶、过筛成茶末，将茶末投入茶盏中，待汤瓶水煮沸后冲入茶盏，再用茶筅不断击拂，使茶汤面浮起一层白色沫饽，茶即点好。宋人对茶汤有特殊的评价标准。斗茶“斗色”“斗浮”，就是比茶汤颜色的白度，以及茶汤表面的沫饽的耐久性。茶汤以纯白为上，青白为次，灰白次之，黄白又次之；沫饽以最后散退、露出水痕的为好。斗茶以茶汤色白、最后露出水痕者为赢。

宋代斗茶成风，上至皇帝，下至黎民，都热衷于斗茶。

960—1279 年

A Nationwide Penchant for Tea Competitions

Tea contest was a custom that measured tea's quality by the color of tea soup and the froth and water mark on the brim.

Though being more subtly processed, cake tea still played the major part during the Song Dynasty. Instead of cooking, the tea-whisking method, or "dian-cha" in Chinese, took center stage. It starts with grinding and sieving for the finest tea powder in place. Gently steep the powder with boiling water for "tea paste", then stir it with a whisk till the white thick froth appears on the mixed brew. Back then, criteria were set to assess tea soup from two angles, i.e, the color of the soup and the lifetime of the white froth. For color, pure white implied the finest quality. Next were the white mixed a shade pale green, gray and yellow. As to froth, concerns lied with for how long it could remain before the water mark appeared. A longer lifetime was called better. Overall, the whisked brew in sheer white and with froth that last dissipated was the champion.

960-1279

For either rulers or plebs, tea contesting was a game superbly in vogue over the Song Dynasty.

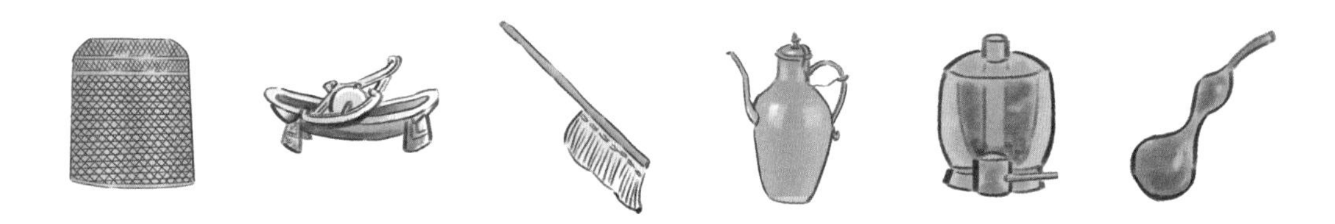

兔毫鹧鸪，建窑黑釉盏自然天成之美

建盏可谓因茶而生，因茶而寂。由于宋代茶色尚白，深色釉的茶盏更能映衬茶汤的白色，便于观察斗茶的水痕，故黑釉盏风行天下。

建窑黑釉盏是宋代黑釉盏中的代表品种。常说的兔毫、油滴、鹧鸪斑即是斑纹不同的建盏。建盏的魅力在于深深的釉色中时隐时现的斑纹，有的细如兔毫、上下通达，有的如油滴点点，有的如鹧鸪的羽毛花纹。建盏的斑纹与盏色全由烧造时的窑温、摆放位置等造就，浑然天成，充满偶然性，因此格外使人着迷。

中国建盏在日本备受珍爱。16 世纪初期的《君台观左右账记》中记载："曜变斑建盏乃无上神品，值万匹绢；油滴斑建盏是第二重宝，值五千匹绢；兔毫盏值三千匹绢。"可见其珍贵。

960—1279 年

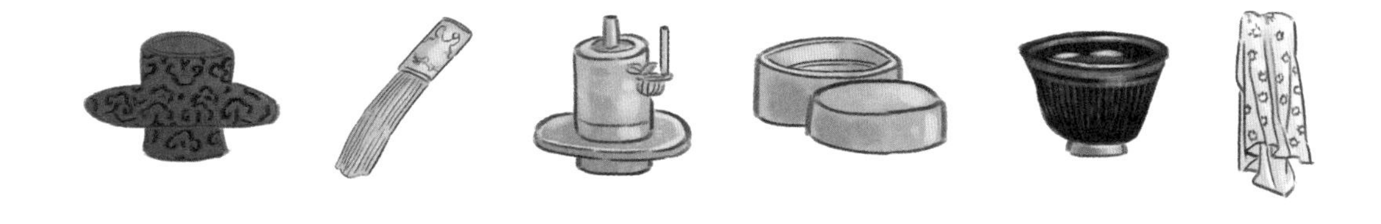

The Ethereal Beauty of Fired Patterns

Nothing but tea could justify the role of "Jianzhan" (pottery of Jianyao Kiln) in China. Mixed tea soup in sheer white was much preferred in the Song Dynasty, while teacups, if painted in dark, could help better define the whiteness and the dissipating water mark. Hence black-glazed teacups started to gain publicity.

The ones from Jianyao Kiln are the most representative sort. Tuhao, Youdi and Zheguban are the names that actually referred to the fired patterns. Specifically, Jianzhan fascinates people with its blurred patterns half-hidden over the dark glaze, representing either rabbit's thin hair, oil drops or partridge's feathers. As those patterns and colors rest completely on the temperature and where the pieces are placed inside the kiln, Jianzhan looks so natural and turns out ethereal collections formed entirely by chance.

Jianzhan is particularly prized in Japan. According to *The General's Collections on Display*, a book completed in the early 16th century, "a priceless cup of 'yao-bian' sort (the pattern of stellar transmutation) costs 10,000 silk sheets, a cup of 'youdi' (oil drop) 5,000 sheets, and 'tuhao' (rabbit hair) 3,000 sheets." They were by then astronomical figures.

960-1279

皇帝茶人，
宋徽宗爱茶著《茶论》

宋徽宗赵佶是一位名副其实的艺术家，他不仅自创了独特的“瘦金体”，绘制了著名的《瑞鹤图》《文会图》等，主持编撰《宣和画谱》，更是中国历史上唯一一个撰写茶书的帝王。他的著作《茶论》成书于大观元年（1107年），故被后人称为《大观茶论》。

1107年

《大观茶论》是一部具有较高价值的技术专著和茶文化经典。全书共二十篇，3000余字，对茶的种植、采制，以及茶具、点茶、品鉴等提出了独到的见解。其中“白茶”记录了茶树变异品种；“点”则非常详尽、细腻地记叙了宋代有代表性的点茶技艺。

宋徽宗著茶书、爱点茶，有力地推动了宋代茶文化的发展。

The Emperor and the Author of a Tea Classic of All Time

Zhao Ji, or Emperor Huizong of the Song Dynasty, was undeniably an artist. He invented the "Slender Gold" style in calligraphy and was the painter of "The Auspicious Cranes" and "The Gracious Party". He sponsored the imperial collection of paintings and was China's only tea book writer in the capacity of emperor. His book *Cha-Lun* (*Treatise on Tea*), completed in AD 1107, the first year of Daguan Period, was therefore widely called *Da-Guan-Cha-Lun*.

Cha-Lun is a masterwork on tea studies and culture. Containing 3,000 Chinese characters over 20 chapters, the book provides Huizong's very insights about tea growing and harvesting, teaware, whisking method, tea appreciation and other concerns. In his book, the emperor recorded the "white tea" as a variant and gave a meticulous account of the "tea-whisking method" so popular across the country, to name a few.

Politician or artist aside, Huizong was assuredly worth a place in the genuine advances of tea culture during the Song Dynasty.

1107

散茶冲泡，
朱元璋引领全新饮茶风尚

因团茶制茶工艺复杂，耗费人力物力，劳民伤财，明太祖朱元璋下诏罢造龙团，唯采芽茶以进。唐宋时期，散茶存在于民间，朱元璋下诏书以散茶为贡茶，这是茶叶采制和品饮方法上一次具有划时代意义的改革。

此后，散茶成为主流茶品，直至今日。散茶大兴，除了使制茶工艺发生变化外，散茶冲泡（又称瀹饮法）取代了延续上千年的烹煮点茶，同时，品饮方式的改变带来了茶具的大变革，碾、罗、筅等废弃，黑釉茶盏也退出历史舞台，取而代之的是白瓷茶具和紫砂茶具。

散茶冲泡的品饮方法带给我们的，除程式的简化外，更多的是茶叶的自然意趣。

1391年

A Call for Loose Tea from the Empire's Founder

Zhu Yuanzhang, the first emperor of the Ming Dynasty, stopped the manufacture of cake-shaped tribute tea that in his eyes was a true waste of money and manpower. Instead, the emperor requested loose-leaf tea. It was back to earlier times, like the Tang and Song, that loose tea was spread over the lower classes. Zhu's decree represented an epoch-making change in the methods of tea harvesting, making and drinking.

Loose tea henceforth took center stage. Changes in tea-making aside, the cooking and whisking methods that had a lifespan of over a thousand years started to give way to the brewing of loose tea. Revolution also took place in utensils. Roller, sieve and whisk were abandoned, and black-glazed pottery cups have left the stage. In their place were white porcelain and purple-clay sets.

It was the brewing of loose tea, a simplified way of drinking, that started to coax more of delights from being closer to the nature.

1391

白瓷紫砂，茶器迎来创新时代

散茶大兴以后，紫砂茶具和白瓷茶具成为茶具的主力军，壶、盏（杯）搭配的茶具组合一直沿用至今。

中国人的陶瓷烧造历史有万年左右，明清以来，瓷器烧造以景德镇为中心。景德镇的白瓷茶具“白如玉，薄如纸，明如镜，声如磬”，能衬托出茶汤的真实色泽，因而成为茶桌上绝对的主角。

宜兴紫砂茶具是陶茶器中的佼佼者，用宜兴的紫砂泥以1000～1200℃高温烧制而成，常见颜色有紫色、红色和黄色。因紫砂泥料透气性好，很适宜泡茶，加之明代以来深受文人喜爱，紫砂壶创作在文人参与下糅合了很多文化艺术元素，除泡茶饮茶外，紫砂壶更可作书房雅玩。

1368—1912年

The Rule of White Porcelain and Purple-Clay Sets

From the nationwide dominance of loose tea came the rule of purple-clay and white porcelain tea sets. The combination of a pot and cups, from that moment, was established and has even remained in vogue today.

It was roughly 10,000 years ago that ceramics first appeared in China. Of all porcelain kilns nationwide, Jingdezhen played the lead particularly from the Ming and Qing Dynasties. Porcelains from Jingdezhen were described as “utterly white as jade, thin as paper, shining as a mirror, and clear in sound like a bell”, and since they could faithfully represent the brew color, they quickly became a blinding star on the table.

1368-1912

Yixing’s purple-clay sets are the pick of Chinese pottery. They are made of zisha, an endemic pottery clay fired at a temperature of 1000 -1200℃ . The fired clay usually appear spurple, red or yellow. With super bair permeability, zisha is an ideal material for tea steeping. So much endeared by the privileged class since the Ming Dynasty, purple-clay pot features the creative mix-and-match of cultural elements in design and can also be a treasurable article in the home.

茶类齐全，制茶工艺日臻成熟

明代制茶工艺有了较大进步。

明初，茶叶杀青由从唐代起占主导地位的“蒸青”变为“炒青”，并大规模应用。炒制工艺使茶叶的香气有所提升。杀青方式的变化让绿茶的制法不断创新。之后，以闷黄工艺为特色的黄茶、后发酵的黑茶和不炒不揉的白茶相继出现。

明朝末期，红茶诞生自福建崇安，名为小种红茶，此后红茶制法陆续传至安徽、江西等地。

清代，以做青工艺为特色的青茶创制，福建崇安、建瓯和安溪等地开始大规模生产。至此，六大茶类均已出现。

1368—1912年

The Finalized Six Tea Categories

Comparatively, the Ming Dynasty saw a significant progress in tea processing.

For "sha-qing" technology, or tea fixation, change took place from "zheng-qing" (steaming) dominating the Tang Dynasty to "chao-qing" (stir fixation) in the early Ming Dynasty. Being of help to intensify tea's fragrance, the stir-and-fry technology began to spread nationwide. More green tea processing techniques were invented with the fixation reform, later creating new categories like yellow tea, post-fermented dark tea and white tea subject to neither stirring nor rolling.

Black tea, or Souchong by the time, was born in Chong'an, Fujian province. Its processing technology later spread to Anhui, Jiangxi and other provinces in China.

Oolong, a new category that takes credit particularly for "zuo-qing" (tossing), was invented during the Qing Dynasty. From this soon came the mass production of oolong tea in Chong'an, Jian'ou and Anxi in Fujian province. All the six major tea categories of China have by then entered the picture.

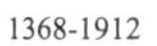

当代“茶圣”吴觉农，《茶树原产地考》论证茶树原产中国

吴觉农（1897—1989年）是中国茶学家、社会活动家，中国当代茶业和茶学的开拓者与奠基人。

吴觉农写《茶树原产地考》，最早论证中国西南地区是世界茶树的原产地，创建中国高等学校第一个茶学系，首创全国性的茶叶研究所，开启科技兴茶之路，负责组建中国茶业公司，主编的《茶经述评》被誉为20世纪的新《茶经》，90岁高龄时，他带头倡导建立中国第一个茶叶博物馆……

吴觉农为中国茶叶事业做出了卓越贡献，被誉为当代“茶圣”。

1897—1989年

吴觉农主编的《茶经述评》
The Classic of Tea Review by Wu Juenong

The Sage of Tea in Modern China

Wu Juenong (1897-1989), a noted tea scientist and social activist, was the groundbreaker of contemporary tea industry and studies in China.

In his paper " *The Origin of Tea Groves*", Wu was the first in the world to prove southwest China the cradle of tea plants. He also took the lead in several areas to restore the magic leaves scientifically to its former prosperity, like China's first tea studies department in university and the first all-China academy for tea research,. He was also the founder of China Tea Company. In his late years, Wu published *The Classic of Tea Review*, a masterwork prized as "the new *Classic of Tea*" in the 20th century, and even at the age of 90, he still urged the building of China's first tea museum.

1897-1989

With all his indelible contributions, Wu is globally revered as the modern "Sage of Tea".

Chapter Two

—

The Classic of Tea

The First, and the Greatest

第二部分

茶的经典

伟大的《茶经》

引文

《茶经》成书于中国唐代（780 年），是世界第一部茶叶专著、第一部茶叶百科全书，被译成日文、英文、俄文、韩文译本，流传于世界。在茶文化史上，《茶经》具有里程碑的意义，对后世茶业与茶文化的发展有着持续、深刻的影响。

《茶经》分为十章：一之源，为概论；二之具，讲制茶工具；三之造，讲茶的采制和鉴别；四之器，讲煮茶与品茶器具；五之煮，讲煮茶和品茶；六之饮，讲陆羽倡导的饮茶方式；七之事，集自神农时期（约公元前 3000 年）至唐代的茶事史料；八之出，讲唐代茶叶产区；九之略，讲茶具可省略的场景和规范；十之图，即建议把整部《茶经》书于白绢并悬挂。

《茶经》不仅奠定了古典茶科学的基本构架，创建了一个较为完整的茶文化学体系，更将“茶”与“经”相提并论，以茶喻君子品行，将茶提升到精神与文化层面。后世茶书创作均未脱离《茶经》的内容框架。

Introduction

The Classic of Tea, completed in AD 780 (the Tang Dynasty), is the world's first comprehensive book on tea studies. Chinese aside, now we have more translations in Japanese, English and Korean. This is a monumental work sweeping the globe with a sustained and profound impact on tea industry and culture.

From ten perspectives (chapters) the author exhausted his learning. First comes the "Origin", an overview of the plant and its morphed existence as a beverage. The next chapter expounds tools for tea picking, processing and storing. "Tea Processing", the third one, details the recommended procedures for tea production and its quality identification. Next comes the "Utensils", elucidating items used in the brewing and drinking of tea. The fifth chapter enumerates the guidelines for proper steps and methods in preparing tea with water from well-chosen sources, while the sixth describes the variegated properties of tea and the author's most-favored drinking methods. The seventh, the "History", gives anecdotes about tea history in literature and records from Shennong's time (circa 3,000 BC) to the Tang Dynasty. It is followed by the "Growing Regions", a chapter that ranks the eight tea-producing regions in Tang China, and next comes the "Omissions" elaborating tools, utensils and brewing methods that can be simplified or improvised under what circumstances. The book winds up with the last called "Reminder", suggesting to transfer the aforesaid contents onto silks and hang them around for quick reference.

The book worked out the framework of the ancient Chinese tea science and a complete system of Chinese tea culture. Defined as a "discipline", tea is compared in this book to a man of honor, elevating itself to an existence of spiritual and cultural significance. No drastic changes have ever been made in all writings following *The Classic of Tea*, either in their thoughts or structure.

《茶经》，
茶圣陆羽书写世界第一茶书

1200多年前，陆羽的《茶经》刻印问世，全书分三卷、十章，仅7000余字，却对世界产生了影响——中国、日本、韩国等国家的茶道思想的形成，乃至后世对茶、水、器的认识，均是在陆羽《茶经》的基础上不断进行的探索、印证和完善。

据传，陆羽是弃婴，被僧人养大，在寺庙苦学12年。754年起，陆羽开始了他的“问茶”之路。他的足迹遍及茶叶产地，考察和总结茶事生产，并将“茶”字作为专属定称；他规范了泡茶器具和饮茶法，提倡“精、俭、和”的茶道精神。

经过26年的编写、修改，780年，世界上最早的茶学经典——《茶经》问世了。

The Classic of Tea: Lu Yu and the World's First Tea Literature

Lu Yu's masterwork first came on the scene over 1,200 years ago. A slim book of 7,000 Chinese characters in three volumes, *The Classic of Tea* was a sensational literature to the world. It was people's unswerving quest, all based on this book, that the thoughts of tea ritual or ceremony took root in China, Japan and South Korea, and that the understanding of tea, water and utensils was shaped and refined.

It was said Lu Yu was abandoned when he was still a baby. He was brought up in a Buddhist temple and spent 12 years practicing prayers and doctrines there. Lu Yu's tea investigation commenced in AD754. He traveled all over the tea-growing areas, poking around for tea farming knowledge and, for the first time, defining this magic plant with the Chinese character "茶", or "tea" in English. He also set standards for tea utensils and drinking methods, and specified the Chinese tea philosophy that "tea is naturally for those who stay modest and are morally disciplined".

It took Lu Yu 26 years to complete the world's earliest tea science literature, which finally came along in AD 780.

《茶经·一之源》

一之源是《茶经》的概论，从“茶者，南方之嘉木也”说起，涉及的内容非常广泛。

首先，陆羽简略介绍了他考察所见的不同品种的茶树，矮小的，仅有30厘米高，高大的，有几米至十几米高，需两人合抱。而后，他用常见的植物做比喻，形象地介绍了茶树和其花、籽、蒂、根的形态。

本章还介绍了茶树生长的土壤、栽培方法和鲜叶品质的鉴别方法、饮茶的益处，以及“茶”字的结构和茶的别名，更以茶喻品行端正、有节俭美德的君子。

The Origin of Tea

"The Origin of Tea", a sketch of the book, starts with tea as "a magnificent tree growing in the South". It straddles a broad range of topics.

Lu Yu first outlined tea trees of different species he saw along the travels. Some were very short, only 30 centimeters high, and some were very tall, ranging from a few meters to a dozen meters high. Tea trees growing to such a size would take two people hand in hand to embrace their circumference. Next, he gave an animated explanation of the shape of tea trees and its flowers, seeds, pedicels and roots by comparing them to common plants.

The first chapter also recounts the soil in which tea grows, skills to tend tea trees, quality identification of fresh tea leaves, benefits of tea-drinking, and even the radicals of "茶" and the four other characters that also denoted it. Tea was also a metaphor in the book for a man of justice and frugality.

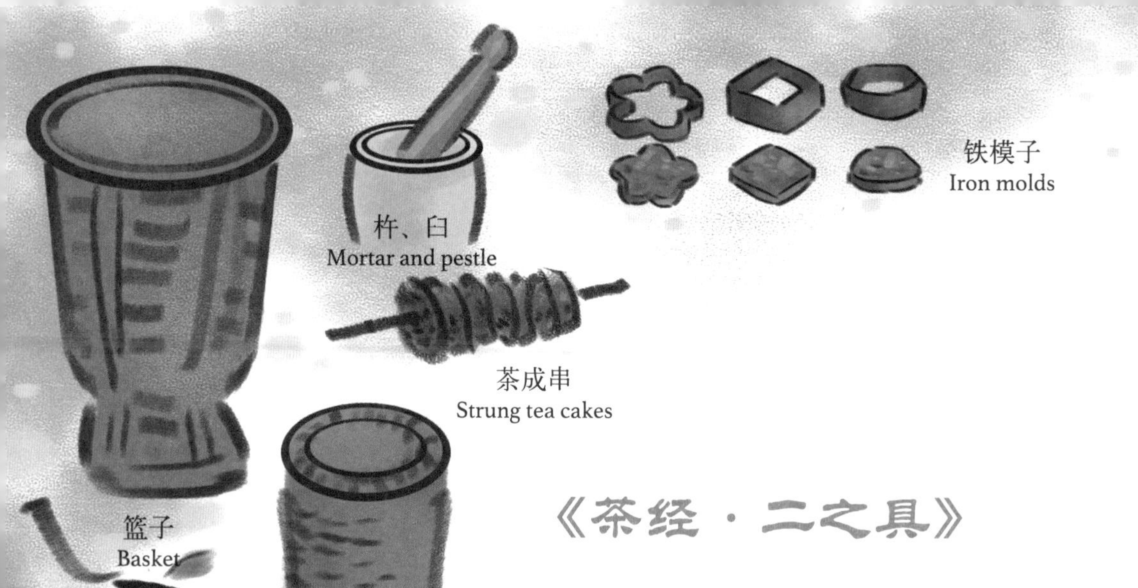

《茶经·二之具》

二之具介绍茶的采制工具。唐代流行的茶叶样式为蒸青饼茶，本章介绍了19种采茶制茶工具：

采茶用的竹篮子；

蒸茶用的炉灶、锅、如同蒸笼的甑、竹蒸隔、木叉；

捣茶用的木杵和石臼；

拍制饼茶的用具，含铁模子、石墩或木墩、垫布、摊开晾干饼茶时用的竹编器具（名芘莉）；

焙茶的工具，包括穿孔的锥刀、穿茶成串的细竹条、烘干饼茶的焙坑和木棚、烘烤饼茶时穿茶用的竹签；

饼茶的计数工具叫穿，一穿五十至一百二十斤；

最后是以木框或竹框糊纸制成的双层烘箱，用于复烘饼茶以去除茶中水分，叫育。

Tools for Tea Processing

Next comes the tools. For steamed cake-shaped tea was in vogue in Tang China, nineteen tools for tea picking, processing and storing were detailed in this chapter.

Bamboo-strip baskets were used in tea picking.

Stove ("*zao*"), wok ("*fu*"), steamer ("*zeng*"), bamboo partition and wooden branch were all used in the steaming of tea.

Wooden mortar ("*chu*") and stone pestle ("*jiu*") were a pair to grind the steamed tea.

Iron mold ("*gui*"), table ("*cheng*") made out of either stone or wood and table cover were used together to press tea leaves into cakes. A bamboo-made device called "*bi-li*", like a sieve, was to hold and dry the separated cakes.

Small awl with a hardwood handle ("*qi*") was employed to punch a hole through each tea cake, while bamboo twine ("*pu*"), in through the holes, was used to string the cakes together. Fire pit ("*pei*") and two-tiered wooden rack ("*peng*"), on which long bamboo skewers ("*guan*") stringing up the cakes were placed, were applied to bake them dry.

Bulk cakes were measured by a unit called "*chuan*", each ranging from 25 kg to 60 kg.

Two-tiered wood (or bamboo) framed storage container ("*yu*"), covered in a paper finish, was the last one mentioned. It was used to re-bake the made cakes and keep them fresh.

《茶经·三之造》

三之造主要讲解了茶的采制方法和品质鉴别方法。

陆羽归纳唐代的饼茶的工艺流程为：采茶—蒸茶—捣茶—拍茶饼—焙干茶—穿起饼茶—封装饼茶。

有趣的是陆羽描述好茶的样子——皱缩不平如胡人的鞋子，皱褶较细像野牛胸部，回转曲折如山间浮云，涟漪荡漾像微风拂水，细腻如水中澄清沉淀的陶泥膏，或如暴雨冲刷的新开垦的土地的，这就是上等精品好茶了。

最后，陆羽又形象生动地描述了茶的品质优劣及原因。

The Processing of Tea

This is a part expounding the procedures for cake tea processing and quality identification.

Lu Yu recapped the procedures by saying the fresh leaves would be steamed, crushed, pressed into cakes, baked dry, skewered and finally sealed.

In his eyes, tea of super quality was in myriad shapes. “Some look like the wrinkles of a barbarian’s leather boots, while others are like the subtle folds of a cow’s neck. They may look like breezy clouds streaming out from behind a mountain peak, or have wavy patterns like the surface of a windswept lake. Some look like clay, fine and malleable, ready to be made into ceramic utensils, while others have the consistency of a field right after ploughing, or the earth after a thunderstorm. These are all signs of prime tea.”

The chapter is wrapped up with tea in various grading scales and an analysis of the reasons behind.

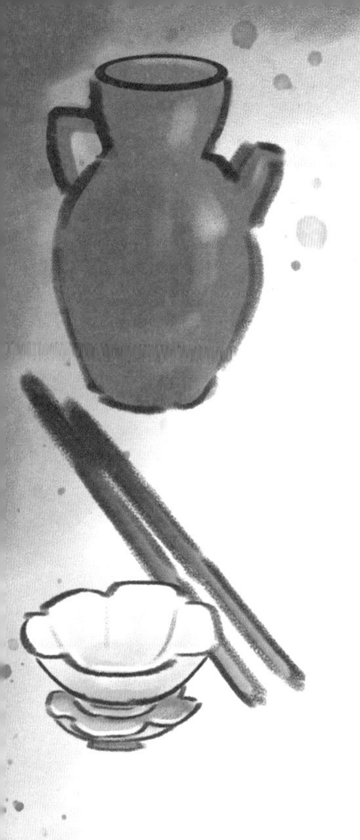

《茶经·四之器》

四之器可谓陆羽给唐朝人开列的一张详细的煮茶、饮茶器物清单，并将每一件器物的材质、制作方法、规格、款式及用途都详细说明。

这些器物大致可分为八类，分别为：

生火器具，含风炉和灰承、竹筥、炭锤、火箸；

备茶器具，含竹夹，纸囊，茶碾和拂末，筛茶的罗、合，量取茶末的茶则；

煮茶器具，含锅、放锅的交床、竹夹；

取水、盛水器具，含盛水用的水方、过滤水的漉水囊、取水的瓢、盛开水用的熟盂；

盛盐、取盐器具，含盐盒和取盐的小勺；

饮茶器具，碗；

清洁用器，含竹刷、盛放洗涤废水的涤方、盛放茶渣的滓方、茶巾；

收纳用器，含盛放茶碗的蒲筐、床形或架形的陈列具、能盛放所有器具的篮子。

这些器物用竹、木、藤、草、贝、布及铜、铁等制成，现代茶具中仍可看到它们的影子。

Utensils for Tea Preparation

The fourth chapter is by nature a list of utensils Lu Yu provided for tea preparing and drinking. Targeting the entire population, the list is also attached with the elaboration of the texture, production, specification, style and use of each item.

All implements elucidated can roughly be divided into the following eight categories.

Furnace ("*feng-lu*"), iron tray ("*hui-cheng*"), hexagonal coal container ("*ju*"), charcoal breaker ("*tan-zhua*") and tongs ("*huo-jia*") are the fire starters.

Bamboo tongs ("*jia*"), paper envelope ("*zhi-nang*"), grinder ("*nian*"), feather brush ("*fu-mo*"), lidded container with a sieve ("*luo-he*") and measuring spoon ("*ze*") are necessary for preliminary processing.

Cauldron or cast-iron kettle ("*fu*"), folding stand ("*jiao-chuang*") and bamboo stick ("*zhu-ce*") are the cooking set.

Water urn ("*shui-fang*"), filter ("*lu-shui-nang*"), ladle ("*piao*") and hot-water basin ("*shou-yu*") are the dedicated receptacles.

Salt dish ("*cuo-gui*") is a round, ceramic salt container with a little spoon.

Bowl is the special item used for tea drinking.

Brush ("*zha*"), waste water container ("*di-fang*"), tea-grounds bin ("*zi-fang*") and tea towel are all the cleaners.

Bowl carrier ("*ben*"), utensil bed or rack ("*ju-lie*") and king-size basket ("*du-lan*") as all-in-one storage are the special teaware vessels.

Apropos of all these antiques, either made of bamboo, wood, rattan, grass, shellfish, cloth, copper or iron, we may still find traces of them in modern tea sets.

《茶经·五之煮》

五之煮论述煮茶的方法，内容包括炙茶、碾茶，燃料的选择、水的选择，水的三沸，煮茶、分茶法以及饮茶法。

陆羽的择水观对后世影响深远。他认为，水以水源分优劣，山泉水为上，江水为中，井水为下，由此引发了后人对评泉鉴水的热衷。

陆羽对煮水“三沸”过程的描写尤为细腻：煮水时，出现鱼眼一样的气泡、微微有声时为一沸，锅边像泉涌连珠时是二沸，水沸腾如波涛翻滚时为三沸。煮茶需在二沸时舀出一勺水，用竹夹搅动沸水中心，投入适量已经碾碎、过筛的茶末，待水沸腾如波涛时倒入刚刚舀出的水止沸、蕴养茶汤精华。分茶时，需每一碗茶汤中的精华——沫（华之薄者曰沫）、饽（厚者曰饽）、花（细轻者曰花）均匀，这样才算煮出一碗好茶！

The Brewing of Tea

Lu Yu recited the brewing of tea in the fifth part. He provided guidelines for the roasting of tea cakes, crushing of the roasted leaves, fuel options, water options, "the three boils" and the suggested methods of tea cooking, serving and drinking.

Lu's grading of water had far-reaching implications. In his point, the source of water counted. "Spring water is the best, river water is average, and well water is not recommended." His comments also led to nationwide obsession to identify the quality and source of water, even today.

The "three boils" were spotlighted. When the boiling water first makes a faint noise and the bubbles are the size of fish eyes, it has reached the first boil ("*yi-fei*"); as strings of pearls arise at the edge of the kettle, it has come to the second stage ("*er-fei*"); if the bubbles grow much bigger and it resounds like drumming, then we have the third boil ("*san-fei*"). During the second boil, we should scoop out a ladle of boiling water for later use. Then, using "*zhu-ce*", the long bamboo stirring stick, to revolve the water at the center of the pot. Next, add measured tea powder with the spoon to the eye of the vortex. Shortly after the tea turns and churns, the water will come to the third boil, roaring like tumbling waves. This is the time to return the hot water spooned out at the second boil to the cauldron. This helps prevent over-boiling and, more importantly, save the tea's essence. It is actually the froth. The thin froth is named "*mo*", the thicker "*bo*", and the light and frail froth "*hua*". In serving tea, we should let the froth settle and spread evenly in the bowls.

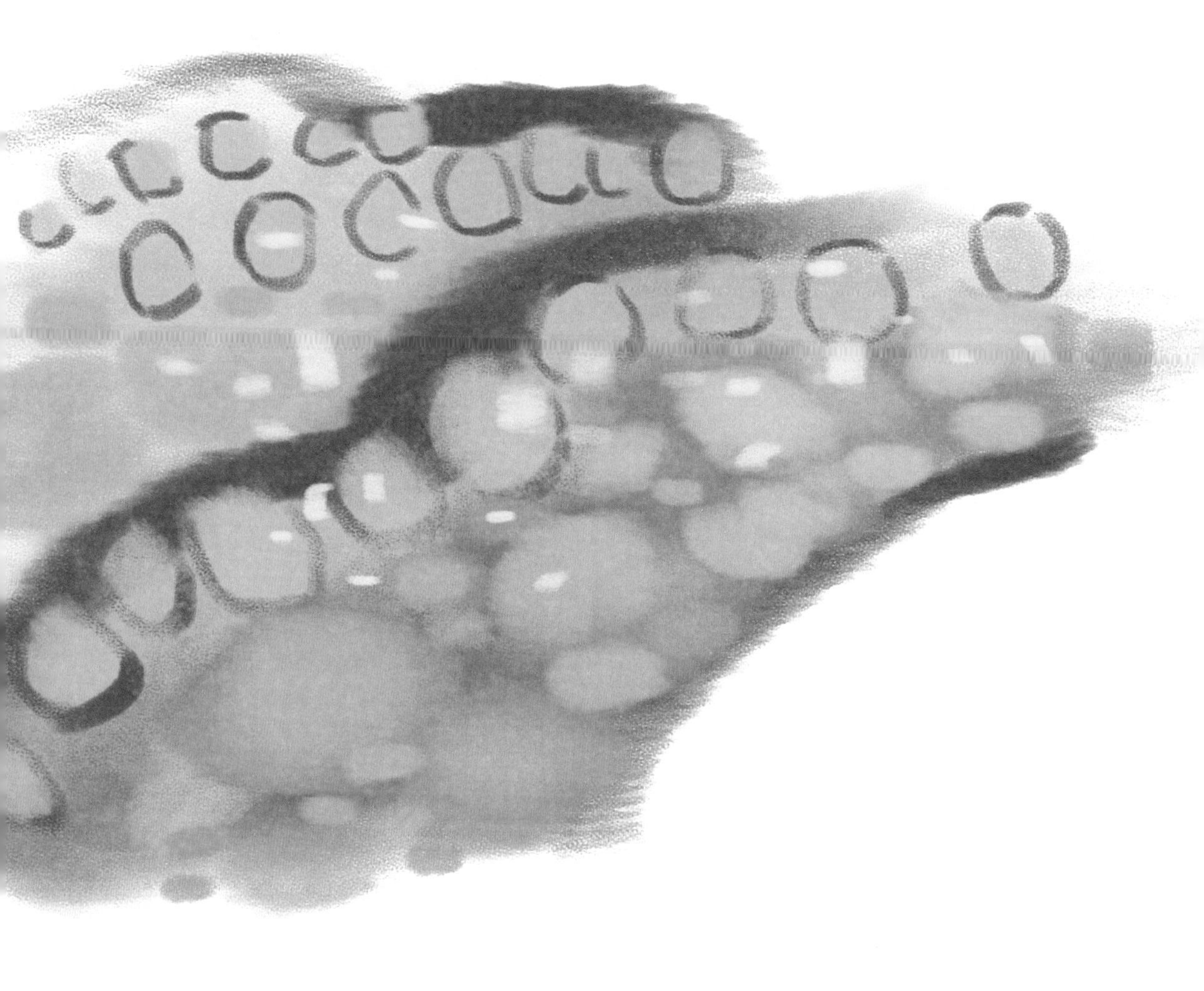

《茶经·六之饮》

六之饮论述了饮茶风尚的传播和陆羽所倡导的饮茶的方式。

“茶之为饮，发乎神农氏，闻于鲁周公”这句话就出自本章，意思是：茶作为一种饮品，源自神农氏，鲁周公时有文字记载。

陆羽认为，民间用开水浸泡茶，或将茶与葱、姜、枣、橘皮、茱萸、薄荷等长时间烹煮，这都是有损茶味的饮茶法。他对茶汤的要求是香气鲜爽、浓郁，滋味深长隽永，并且一茶则的茶末，只应煮出三碗茶。他罗列了茶的九个难点、不能用于煮茶的器物和水等细节，这些都体现了陆羽对待茶饮的观点——技术要“精”，饮茶重“品”，更加注重饮茶带来的精神享受。

The Drinking of Tea

The sixth part, called "Drinking", records the spread of tea customs and explains the way of drinking Lu Yu himself endorsed.

It was from this chapter that came the much-cited saying: "now tea as a drink was first discovered by Shennong, and first recorded by Zhougong (Duke of Zhou)."

In Lu Yu's eyes, it was absolutely wrong of simply steeping fragmented tea with hot water in a bottle or cooking tea long with green onions, ginger, dates, orange peels, dogwood and/or mint, as many people did. He believed it would wreck the true flavor. Instead, he expected the brew aromatic, strong and mellow. For a spoonful of tea powder, the essence should manifest in but three bowls. He enumerated nine skills or "not-to-dos" significant to a life of tea, covering utensils and water ill-suited to the brewing. They all reflected Lu's stands that brewing techniques mattered and tea was an elixir to "deliberately taste", and that in sips, the tea-evoked pleasure in mind should prevail.

《茶经·七之事》

七之事较全面地收录了从上古至陆羽生活的唐代，有关茶的历史资料48则，内容可分为医药、史料、诗词歌赋、神异、注释、地理和其他等七类。这在当时资料难以获得的情况下，是一件很了不起的事情。

七之事中收录了很多现今人耳熟能详的茶故事，如广陵老姥卖茶救济贫苦百姓的故事，韦曜“以茶代酒”的典故，陆纳倡导“以茶养廉”的故事等。此外，关于茶的功效，从本章收集的古籍资料中可看出，古人认为茶能解毒、健体、益思、驱睡、消食。

此外，本章还留下茶叶生产与茶叶贸易、尚茶之风的普及等史料线索，为后人研究古代茶文化留下了宝贵资料。

The Historical References

Forty-eight anecdotes, dating from time immemorial to the Tang Dynasty when Lu Yu was alive, are given from the angles of medicine, history, literature, legend, commentary, geography and others. This is quite an accomplishment considering the scarcity of written black-and-white at the author's time.

Stories now familiar to everyone are also in the collection. The old lady giving all she earned from her tea business to the homeless, Wei Yao's privilege of drinking tea instead of liquor before the emperor, and the anecdote of Lu Na offering tea and fruits to a dignitary are just a few to name. The author also mentioned the efficacy of tea. We believe, from what the chapter says, that tea was considered in ancient China an antidote and good for health, and an herbal drink that helped keep mind alert, dispell drowsiness and digest food.

Besides, references as regards tea production and trade, and the dissemination of tea customs as well, are priceless in ancient Chinese tea culture studies.

《茶经·八之出》

八之出介绍了陆羽所知的唐代的茶叶产地，以及各地所出产茶叶品质的等次。

陆羽根据考察、资料收集，把他所知道的茶叶产区汇总为山南、江南、淮南、剑南、岭南、黔中、浙西、浙东8个道，有的还详细到某山、某村、某寺。这一区域大致为现在的湖北、湖南、陕西、河南、安徽、浙江、江苏、四川、贵州、江西、福建、广东、广西等13个省（自治区）。足见唐代的茶叶产区已相当广大。

陆羽还列出各茶产地出产的茶叶及品质，如：浙西茶区以湖州所产的茶品质最好，湖州茶产于顾渚山谷。可见，现在很多名茶在唐代就颇有名气了。

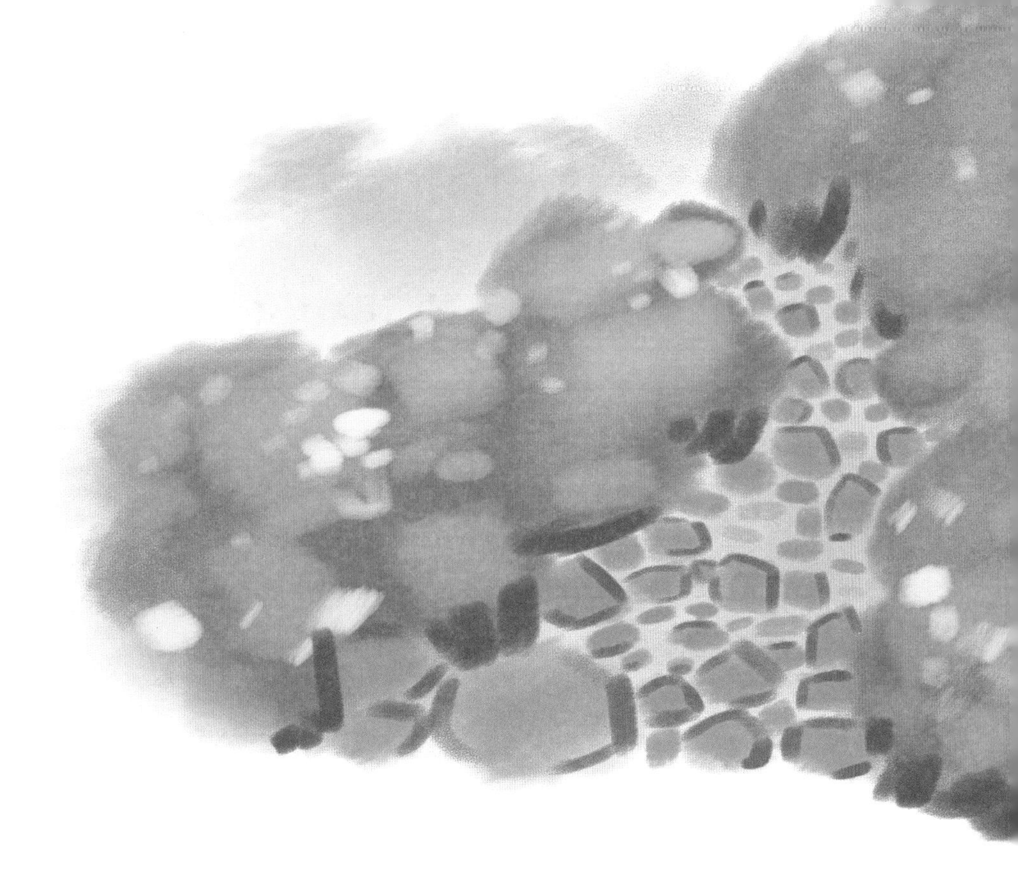

The Growing Regions Across China

In the eighth chapter, Lu Yu named within his knowledgethe tea-growing areas across China and the grading of teas from them.

All these areas, in line with the author's hands-on investigations and data collection, are summarized into circuits of Shan'nan, Jiangnan, Huainan, Jian'nan, Lingnan, Qianzhong, Zhexi and Zhedong, sometimes even with the name of a hill, village or temple. His map covers roughly thirteen provinces (or autonomous regions) today, including Hubei, Hunan, Shaanxi, Henan, Anhui, Zhejiang, Jiangsu, Sichuan, Guizhou, Jiangxi, Fujian, Guangdong and Guangxi—the immensity of Tang's tea producing territory was much in evidence.

Teas from the said areas and their quality are also registered. In the "tea belt" of west Zhejiang, for example, tea from Huzhou, or technically from the valleys of Mt. Guzhu, is the best. Loads of teas we are now familiar with, thus in all probability, were quite in demand back to the Tang Dynasty.

《茶经·九之略》及《茶经·十之图》

九之略，可理解为可以省去的工具、器物，以及省略的规范。陆羽认为，在特定时间、地点等客观条件下，有的工具和器皿可以省略。比如在松间煮茶，石上可放置茶器，则陈列器具可以省略；如用干柴和锅煮茶，则风炉、灰承、火夹等器具可以省略；如在泉边溪畔煮茶，则取水、盛水器具都可以省略……但是，如果是在城里王公之家，二十四件茶器则缺一不可，缺少一件，就不是真正的茶饮了。

十之图，是建议把整部《茶经》都书写在白绢上，悬挂于座位旁，方便随时观看和记忆《茶经》的内容。

陆羽著《茶经》，是希望他对待茶饮的初衷能被人理解，“精行俭德”的茶道理念和饮茶法得以流传。

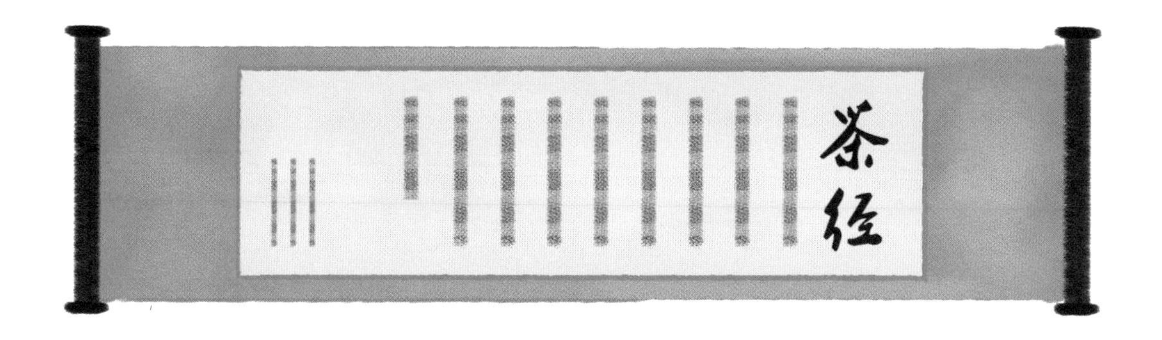

Omissions and the Text Reminder

The ninth chapter stands for any implements or utensils, and under what circumstances in tea processing, to be left out. Such circumstances, like what time and place, were acceptable in Lu Yu's theory. If one brews tea in a pine forest and happens upon stones large enough for him to sit upon, then a utensil rack will not be necessary; if one uses dried firewood and a tripod, there is no need to bring the brazier, ash tray, fire tongs etc.; and if one brews tea along the riverside where fresh water is at hand, then the vessels for drawing and storing water can all be left behind. However, whenever brewing tea far from the wilds, within of the gates of the aristocrats, all twenty-four utensils and tools are needed to prepare the finest tea. The entire process would fail if any one of them is missing.

The last one, or simply "the Reminder", calls for the transfer of the whole book onto white silks and hang them around by the seat. It provides a quick access to learning and memorizing the texts.

In Lu Yu's anticipation, his book could be of any help to people's understanding of his own thoughts about tea. It would also be a monument to Lu's philosophy defining tea as "a token of modesty and moral disciplines" and its processing and drinking methods he preached all his life.

Chapter Three

—

The Knowledge of Tea

A Beverage for the Global Village

第三部分

茶的品赏

天南海北一壶春

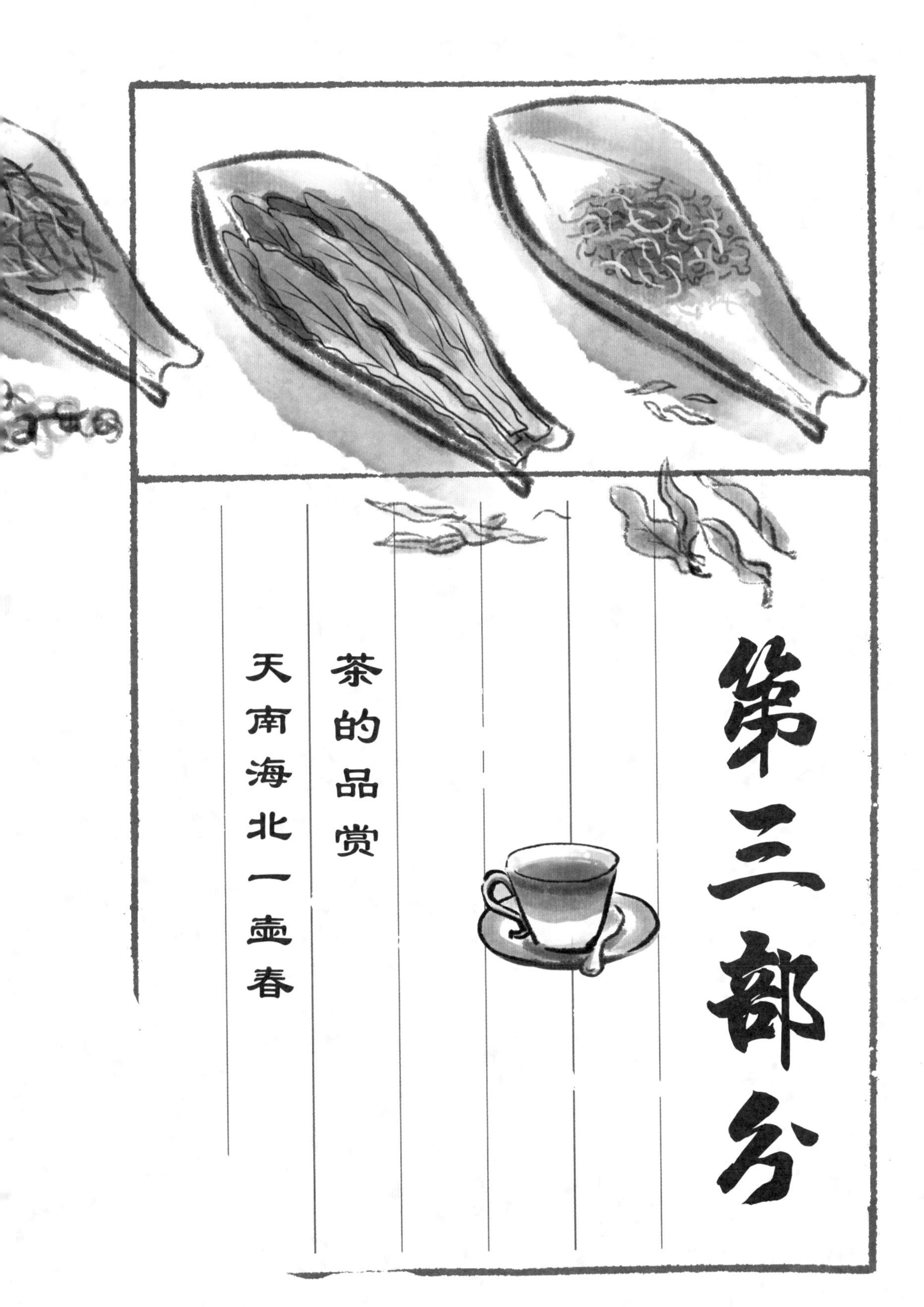

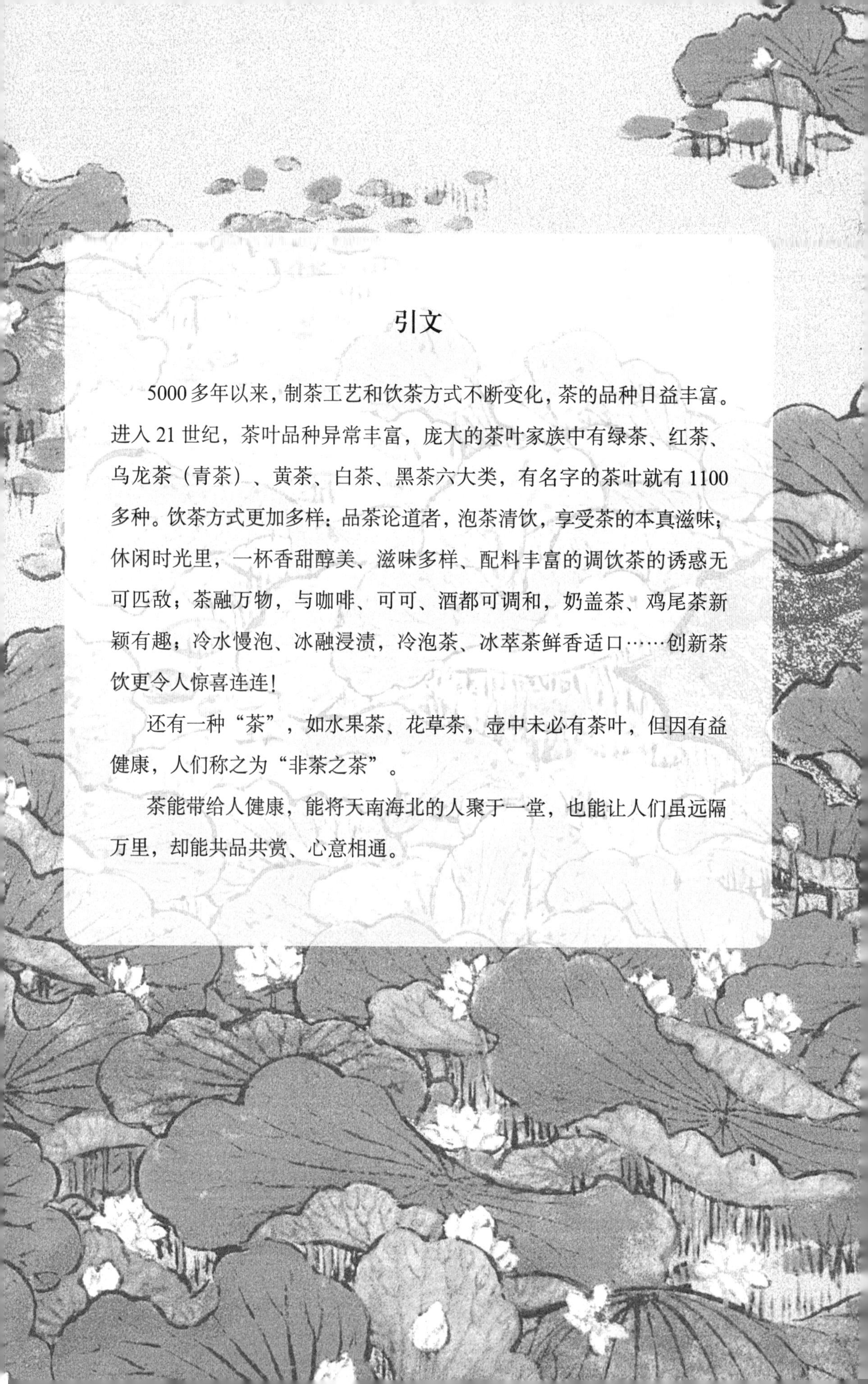

引文

5000多年以来，制茶工艺和饮茶方式不断变化，茶的品种日益丰富。进入21世纪，茶叶品种异常丰富，庞大的茶叶家族中有绿茶、红茶、乌龙茶（青茶）、黄茶、白茶、黑茶六大类，有名字的茶叶就有1100多种。饮茶方式更加多样：品茶论道者，泡茶清饮，享受茶的本真滋味；休闲时光里，一杯香甜醇美、滋味多样、配料丰富的调饮茶的诱惑无可匹敌；茶融万物，与咖啡、可可、酒都可调和，奶盖茶、鸡尾茶新颖有趣；冷水慢泡、冰融浸渍，冷泡茶、冰萃茶鲜香适口……创新茶饮更令人惊喜连连！

还有一种“茶”，如水果茶、花草茶，壶中未必有茶叶，但因有益健康，人们称之为“非茶之茶”。

茶能带给人健康，能将天南海北的人聚于一堂，也能让人们虽远隔万里，却能共品共赏、心意相通。

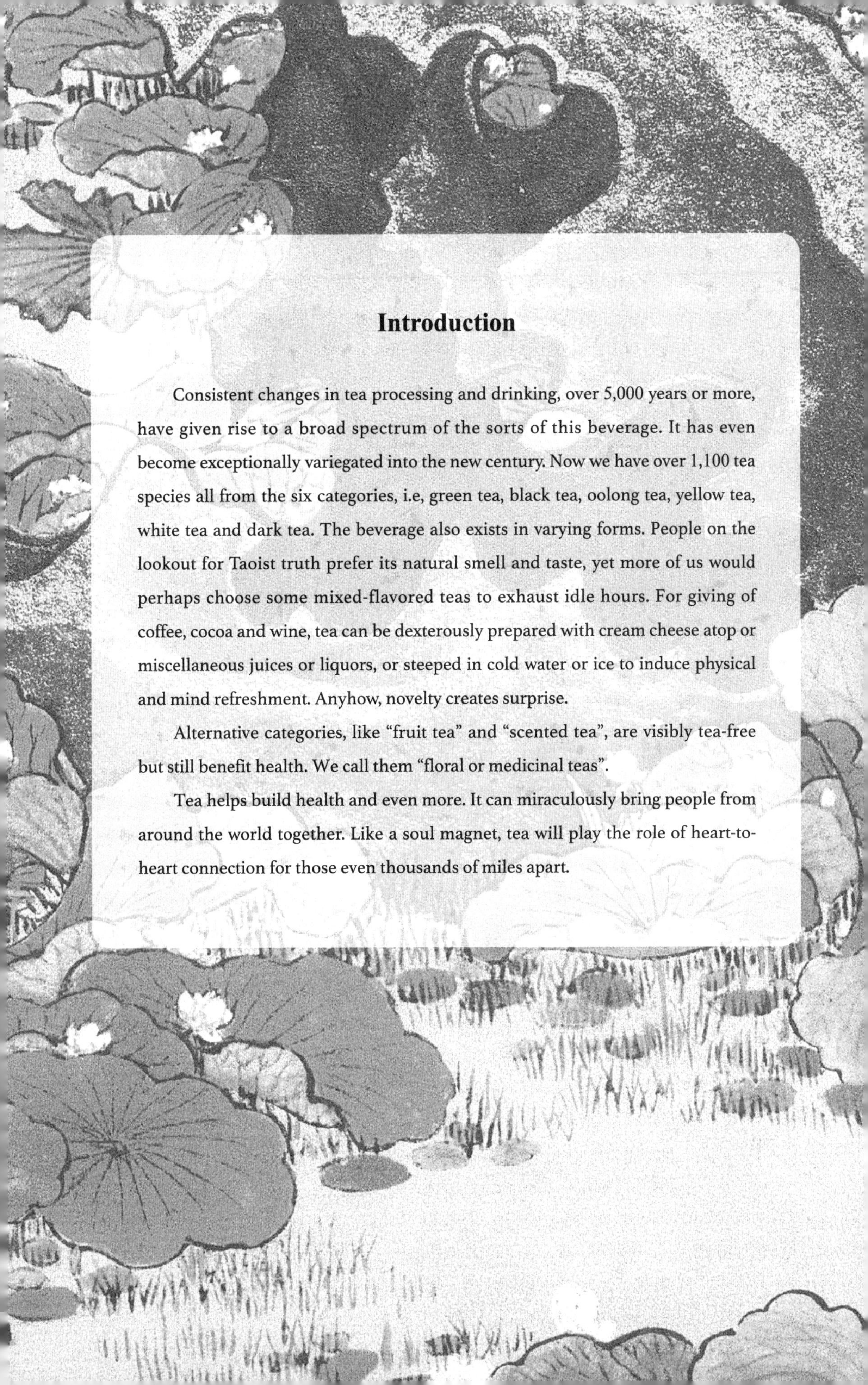

Introduction

Consistent changes in tea processing and drinking, over 5,000 years or more, have given rise to a broad spectrum of the sorts of this beverage. It has even become exceptionally variegated into the new century. Now we have over 1,100 tea species all from the six categories, i.e, green tea, black tea, oolong tea, yellow tea, white tea and dark tea. The beverage also exists in varying forms. People on the lookout for Taoist truth prefer its natural smell and taste, yet more of us would perhaps choose some mixed-flavored teas to exhaust idle hours. For giving of coffee, cocoa and wine, tea can be dexterously prepared with cream cheese atop or miscellaneous juices or liquors, or steeped in cold water or ice to induce physical and mind refreshment. Anyhow, novelty creates surprise.

Alternative categories, like "fruit tea" and "scented tea", are visibly tea-free but still benefit health. We call them "floral or medicinal teas".

Tea helps build health and even more. It can miraculously bring people from around the world together. Like a soul magnet, tea will play the role of heart-to-heart connection for those even thousands of miles apart.

红茶，融融的暖意

400多年前，崇安县（今福建武夷山）桐木关首创小种红茶，因此，中国武夷山桐木关被称为红茶的发源地。红茶是最早走出中国国门的茶类。1610年，红茶由海路运往荷兰，之后相继运至英国、法国和德国等地，让那里的人们爱上了红茶。目前，红茶是全球产量最高的茶类。

红茶是发酵茶，发酵是加工的核心工序，造就了红茶“红汤红叶”的品质特征。红茶茶汤颜色红明，叶底红亮，散发着令人愉悦的甜香、花果香。易于融合的特点使红茶变换出百般风味——与糖、奶及奶制品、蜂蜜、果干等都能完美搭配，红浓醇香又明艳动人。用一杯暖暖的红茶唤醒自己，体贴而温暖。

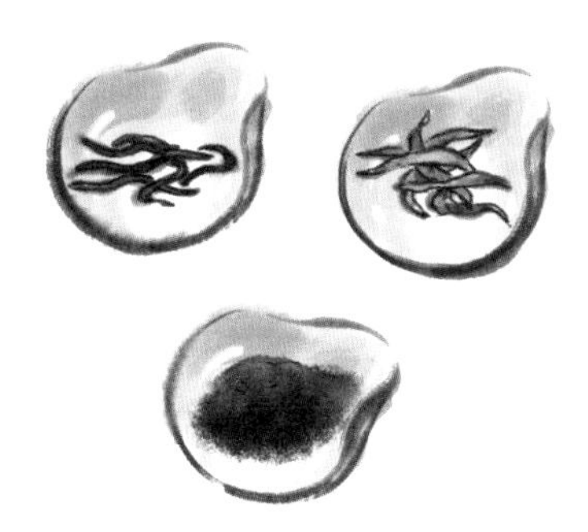

Black Tea: A Must in Chilly Days

Souchong, a fine black tea variant, was first made four centuries ago at Tongmuguan of Chong'an county (now Wuyishan, Fujian province). The place is therefore reckoned the birthplace of black tea. Of China's six tea types, black tea was the first to be unveiled to the world. Its maiden voyage ended up with the Netherlands in 1610, and next the Britain, France and Germany. The exports soon became in vogue along the seaway. Today, black tea takes the most share in global tea farms.

Black tea falls under the category of fermented teas. It is fermenting as the core of techniques that induces black tea's glossy, crimson soup and post-brewing leaves inside, usually exuding a sweet, pleasant fruit-like aroma. Black tea seems by nature all-embracing in a mix with any of ingredients, like sugar, milk, dairy, honey or dried fruits, and each bearing an outstanding flavor. What to do for a good day? A sip of hot, strong black tea will be the answer.

绿茶，萌动的春意

绿茶是中国产量、消费量最大的一类茶，全国 20 个产茶省（区）都生产绿茶。中国绿茶种类居世界之冠，出口量世界第一，每年达数十万吨之多，占世界绿茶贸易量的 70%左右。

绿茶是不发酵茶，因此较多地保留了鲜叶的天然物质，具有干茶绿、茶汤绿、叶底绿的“三绿”典型特征。早春绿茶最为珍贵，茶树经过冬天的蓄养，早春萌出一颗颗嫩芽，茶芽被采下，经历摊放、杀青、揉捻、干燥工艺，成为扁形、针形、螺形、条形、雀舌形、片形等形态的干茶，再被沸水唤醒，在杯中吸水绽放，终成一杯汤清叶绿的绿茶，散发出清香、栗香或嫩香，滋味鲜爽怡人。

Green Tea: A Messenger of Spring

Green tea tops the rankings of production and consumption in China. As the category earns a place in the country's all 20 tea-growing provinces (or autonomous regions), China outperforms any parts of the world in green tea sorts and export. The aggregate, up to hundreds of thousands of tons annually, accounts for 70% of the world's green tea transactions.

"Unfermented" in nature, green tea retains most of the natural ingredients in fresh leaves. Conventionally, it manifests a rich green either in dried form or in the forms of brew and the post-brewing leaves. The pick of green teas come in early spring, a time of sprouting buds to part from the last winter. The fresh-picked buds are amassed for a process that spans spreading, fixing, rolling and drying before in flat, needle, spiral, strip, bird-tongue or sheet-like shapes. Steeped with boiling water, the leaves are gently brought to life. The finalized being, with its brew and leaves both in adorable green, gives off a fragrant smell sometimes like chestnut or new grass. It tastes mellow and fresh in the main.

青茶，多变的气质

青茶又叫乌龙茶，属于半发酵茶，介于不发酵的绿茶与全发酵的红茶之间，兼具绿茶的清香和红茶的醇厚。它色泽青褐，所以称之为“青茶”，又因冲泡后叶片中间呈绿色、边缘为红色，故称“绿叶红镶边”。

青茶是加工工艺较为复杂的茶类，也是具有馥郁天然花果香的茶类。“做青”是形成青茶丰富香气和醇厚滋味的关键工序。由于做青、发酵和焙火的程度轻重不同，青茶的香气可谓千变万化——韵味独特醇厚的大红袍、似桂皮香或乳香的肉桂、熟果香和蜜香怡人的东方美人、兰香持久的铁观音、花果香清雅的冻顶乌龙、花香清高汤色金黄的凤凰单丛……它们都是青茶。青茶外形主要有两种，包揉形成重实的颗粒形，以及揉捻成粗壮的条索形。冲泡青茶适宜使用小壶，用滚烫的沸水唤醒它的香气和滋味，回味无穷。

Oolong Tea: A Lady of One Hundred Faces

Oolong, a partially-oxidized type that boasts a full-bodied, floral smell, falls between unoxidized green teas and fully-oxidized black teas. As the dry leaves usually appeara mix of blue and green, oolong is also termed "blue-green tea" ("*qing-cha*"). The leaves will turn green with a red trim in very hot water, an arresting phenomenon called "red-rimmed green body".

The complex processing also fills oolong tea with a lingering smell of flowers or fruits. For creating such an incredible smell and taste, a step called "bruising (*zuo-qing*)" plays foremost. The smell of oolong varies noticeably with the variances of its bruising, fermenting and baking degrees. "Da-Hong-Pao" (with a strong, mellow taste), "Rou-Gui" (exceptionally with a cinnamon/milk-like flavor), "Dong-Fang-Mei-Ren" (with a delightful smell of ripe fruits or honey), "Tie-Guan-Yin" (with an unabating orchid-like flavor), "Dong-Ding Oolong" (with a graceful, nutty smell and taste) and "Feng-Huang-Dan-Cong" (with a floral smell and a golden soup)..., to name just a few. Tight or soft rolling gives oolong two different looks in general, i.e, like semi-nuggets or long twists. It is more favorably prepared in small teapot, and should go with boiling water to give full rein to its rich, abiding aroma and taste.

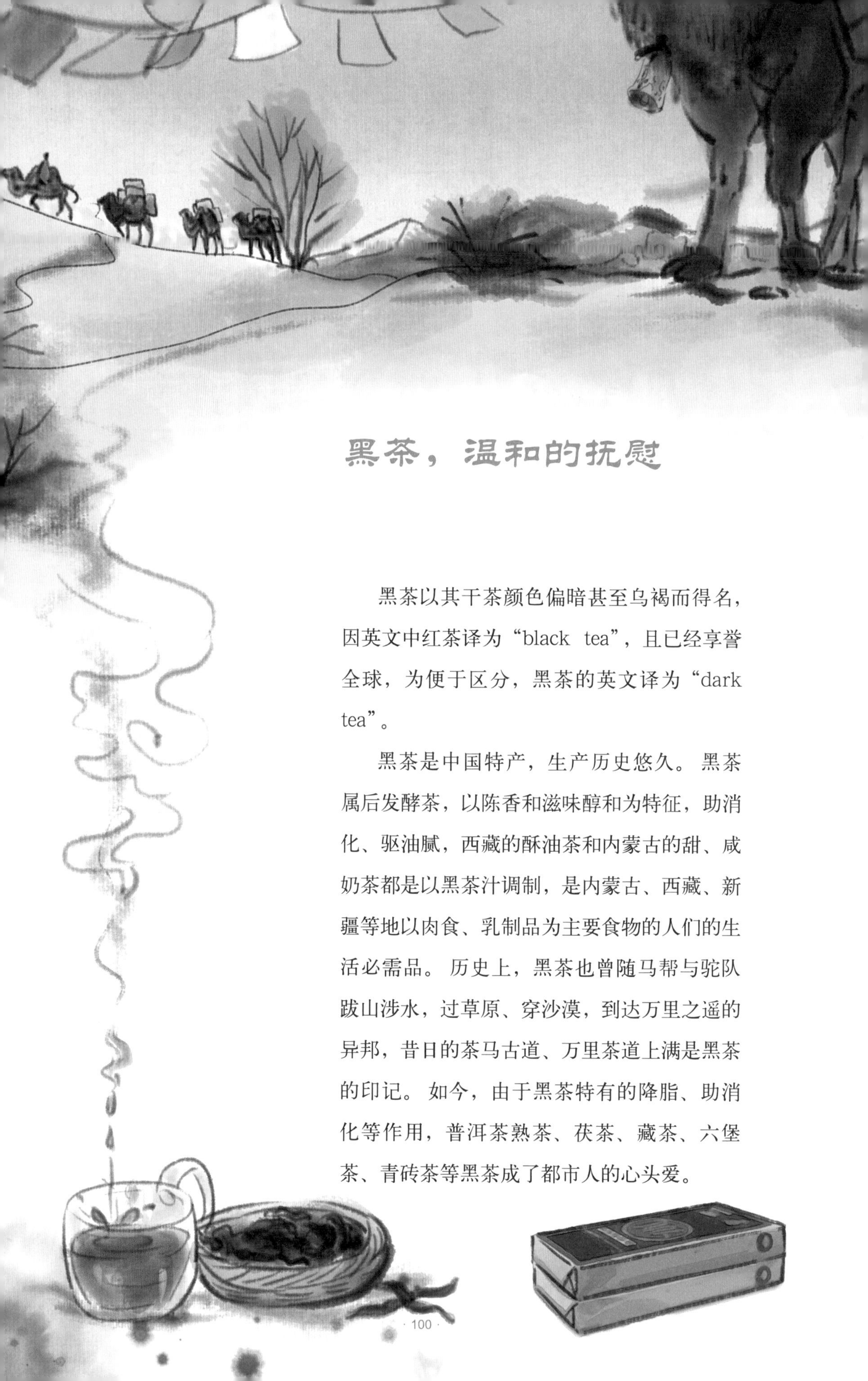

黑茶，温和的抚慰

黑茶以其干茶颜色偏暗甚至乌褐而得名，因英文中红茶译为“black tea”，且已经享誉全球，为便于区分，黑茶的英文译为“dark tea”。

黑茶是中国特产，生产历史悠久。黑茶属后发酵茶，以陈香和滋味醇和为特征，助消化、驱油腻，西藏的酥油茶和内蒙古的甜、咸奶茶都是以黑茶汁调制，是内蒙古、西藏、新疆等地以肉食、乳制品为主要食物的人们的生活必需品。历史上，黑茶也曾随马帮与驼队跋山涉水，过草原、穿沙漠，到达万里之遥的异邦，昔日的茶马古道、万里茶道上满是黑茶的印记。如今，由于黑茶特有的降脂、助消化等作用，普洱茶熟茶、茯茶、藏茶、六堡茶、青砖茶等黑茶成了都市人的心头爱。

Dark Tea: A Panacea in Life Diet

"*Hei-cha*" literally means "black tea" in Chinese. Considering the name universally refers to another category to extract red/orange soup, "*hei-cha*" is finally called "dark tea" for its dried leaves in varying auburn or dark brown shades.

This type is endemic only to China. Over a prolonged history, dark tea undergoing microbial fermentation (post-fermented)has featuredparticularly an "aged" fragrance and a mellow taste. As it helps with digestion and neutralizes greasy foods, buttered tea from Tibet and sweet/salty milk tea from Inner Mongolia are both prepared with dark tea brew. This is a life essential across Tibet, Inner Mongolia and Xinjiang, where meat and milk are stable foods. Centuries ago, dark tea was made exports travelling on horses or camels over the mighty wild grasses and deserts, a trek spanning thousands of milesto either southeast Asia or Russia and around. Now a blood-fat killer and a digestive aid, esteemed dark teas, like "Pu'er", "Fu-Cha", "Zang-Cha", "Liu-Bao" and "Qing-Zhuan", are popular sellers in all major cities of China.

黄茶，茶中的小众

黄茶也是中国特产，其产量和品种都不多，因此说它是茶叶中的“小众”。黄茶“黄叶黄汤”，干茶嫩黄、黄青或黄褐色，茶汤黄亮，香气有清香或锅巴香，滋味醇和，有回甘。

黄茶属轻发酵茶，其创制或许是制茶匠人的无心之举。黄茶的加工工艺与绿茶基本相似，只是增加了一个闷黄工序，仅仅这样一“闷”，茶色黄、汤黄，茶香清悦，比绿茶少了鲜爽，多了醇和。

黄茶按原料芽叶的大小和嫩度分为黄芽茶、黄小茶和黄大茶三类。黄芽茶的冲泡特别富有观赏性，杯中的茶芽悬空竖立，徐徐浮沉，如茶芽在杯中起舞。

Yellow Tea: A Special for Niche Markets

Yellow tea is also a native category only to China. For a dearth of supply and sorts, it seems like a minor tea species. The prepared soup turns purely yellow, so do the steeped leaves. It looks fresh yellow, chartreuse or tawny in dried form. A steeped yellow tea smells flowery or close to rice crust, with a rich, mellow taste and a sweet aftertaste.

Yellow tea was probably a chance discovery. Similar to making green tea in all steps, the processing of yellow tea, a lightly-fermented category, applies a new tactic simply called "yellowing". It is precisely the said extra step that yellows the soup and post-brewing leaves. Compared to green tea that tastes fresh, yellow tea prioritizes a full-bodied, mellow flavor.

The family of yellow tea, varying in the size and freshness of raw leaves, consists of "Huang-Ya-Cha" (yellow buds), "Huang-Xiao-Cha" (small yellows) and "Huang-Da-Cha" (big yellows). Brewing a "Huang-Ya-Cha" seems particularly a close-up magic. The yellow buds, as in hot water, turn upright, uncertainly hovering like an aerial dance.

白茶，不饰的素颜

白茶为中国特有的茶中珍品，因其干茶外观灰白或披满白毫，茶汤色浅而得名。依据原料嫩度和茶树品种不同，白茶分为白毫银针、白牡丹、贡眉和寿眉，茶类名即茶叶名。

白茶属微发酵茶，是加工工艺最精简的一种茶类。茶鲜叶采下后萎凋、干燥，不炒不揉，干茶外形舒展，自然天成，尤其是白毫银针，一颗颗绿芽银毫，色泽银亮，肥壮饱满，呈针形，堆放的白毫银针远看就像一座白雪皑皑的小山，素颜不饰，清雅从容。

白茶汤色清亮，香气清鲜，滋味鲜醇。白茶当年采制当年饮用是一种风味，存放几年后饮用又有与新茶不同的风味，故有“一年茶，三年好，七年宝”之说。

White Tea: A Gem Preferably Made Natural

High-priced a category unique to China, white tea is so named for its soup looks subtly colored, and the dry leaves usually appear pale grey or thick with white trichomes. The entire stocks are usually sorted into "Bai-Hao-Yin-Zhen" (silver needles), "Bai-Mu-Dan" (white peony), "Gong-Mei" and "Shou-Mei" as per leaf freshness and floral species. The species names also apply to the brand names.

The slightly fermented category allows a stripped-down processing tactic. Leaves picked fresh are not to be fried or rolled. Through extended withering and drying, the leaves are unfolded as in wild nature. The dried leaves of "Bai-Hao-Yin-Zhen", for example, are stout and needle-like, coated with gleaming white trichomes. The sort in stacks, from a distance, looks like a snow-capped hill. Free of any makeup, but it is still in fair shape.

In brewing white tea, the soup usually appears bright-colored. It smells fresh and tastes vaguely sweet with floral notes. A white tea prepared the same year it was picked pledges a flavor much different from what's given by tea stocked for years. Therefore, we have a saying that "white tea aged less than a year is average. It is good if aged three years, and priceless if seven years".

花茶，茶与花的交融

用加工好的茶叶，如绿茶、红茶、乌龙茶等，与茉莉花、玫瑰花、桂花、白兰花、代代花等鲜花混合，让茶叶充分吸收花的香气，制成的茶叶带有馥郁的花香，冲泡后花香与茶香沁人心脾，令人仿佛置身花海，身心安宁，妙不可言。这就是中国特有的再加工茶——花茶，常见品种如茉莉花茶、玫瑰红茶、桂花乌龙等。

如果一个中国人说他喜欢喝花茶，那“花茶”一定是指用绿茶和茉莉花制成的茉莉花茶。茉莉花香气浓郁而清雅，被誉为“人间第一香”，茉莉花茶的主要特征自然是“香”，其主要工艺为“窨制”，即将茶坯与刚采摘的含苞待放的茉莉鲜花混合在一起，鲜花吐香，茶吸花香。优质茉莉花茶香气鲜灵浓郁、汤色黄亮明净、滋味鲜爽浓醇，沏一杯花茶，满屋的香气瞬间令人感到轻松愉悦与安宁。

Floral Tea: A Mix of Fragrance and Flowers

The re-processed floral tea (*hua-cha*), a special unique to China, refers to green, black or oolong tea mingled with scented flowers (like jasmine, rose, osmanthus, magnolia, bitter orange etc.). The total absorption lends it a bold floral smell and its brew a bracing mixed fragrance, like a panacea for inner peace. Jasmine tea, rose (black) tea, sweet-osmanthus (oolong) tea are the common few in China.

Jasmine tea epitomizes the mix of floral and tea fragrances. It is made by scenting dry tea leaves with fresh jasmine buds. The concept of "scented tea", or "floral tea" ("*hua-cha*"), refers to the mix of fine green tea and fresh-opened jasmine blossoms.

Jasmine is particularly a natural fit with green tea for Chinese. For jasmine tea, the floral scent comes essentially from the flower with "a most-favored sweet fragrance". "Scenting", a core step in making jasmine tea, refers to a process in which fresh jasmines are stored with crude green tea absorbing the flowers' smell. A good jasminc tca bears a bold, perfume-like scent and breeds bright, yellow soup. It tastes full-flavored, instantly filling the room with a soft, calming fragrance.

功能茶，非茶的“茶饮”

中国人习惯将有益健康的植物泡饮，也称之为“茶”，如菊花茶、苦丁茶、桑叶茶、藤茶，以及决明子茶、西洋参茶等。可以说，所有用植物的根、花、叶、果实等加工成的对健康有益的饮品，我们都称之为“茶”，习惯上被归入“非茶之茶”。

饮用花草茶、果干茶等非茶之茶并不是中国人独有的习惯，世界上很多地方都有将身边的一些植物加工后泡饮的传统，如南美洲很多地方的人喜欢饮用马黛茶，欧洲国家很早就流行饮用百里香、马鞭草、鼠尾草等花草茶，亚洲地区则流行饮大麦茶、苦荞茶、玄米茶、柚子茶等。

非茶之茶普遍存在于世界各地，体现了劳动人民的生活智慧。

Herbal Tea: A Beverage with Non-tea Herbs

All for-health beverages done with herbs are named "tea" in China, including chrysanthemum tea, kuding tea, mulberry tea, rattan tea, cassia seed tea and American ginseng tea, to name a few. In other words, any "tea" actually using the roots, flowers, leaves and fruits of any floral species fall into the category of herbal teas if good for health.

The consumption of herbal or dried fruit tea is not unique to China. It has been a worldwide practice to have floral parts processed and steeped in water or liquor. For example, yerba mate has been a popular drink in most of South America, and herbal teas like thyme, verbena and sage have remained in fashion across Europe since early on. For Asia, however, barley tea, buckwheat tea, brown rice tea and pomelo tea are the most accepted "tea drinks".

Herbal teas now prevalent all over the world are the fruits of everyday life.

清饮茶，品味茶的本真况味

清饮，是指茶汤中不添加任何其他调味品、饮品和食物，享受茶的原汁原味的茶饮。自唐代陆羽在《茶经》中倡导清饮以后，清饮一直是中国人的主要饮茶方式。潮州工夫茶是最具特色的乌龙茶清饮茶艺。

中国是茶的故乡，茶树种质资源丰富，制茶工艺多样，泡茶饮茶追求精妙，而清饮最能品味茶树品种特色、加工工艺以及产茶地的那方水土共同赋予茶的特殊况味。

无论是中国六大茶类名品迥异的韵味，还是世界著名高香红茶的迷人特质——中国祁门工夫红茶似蜜糖、似兰花的“祁门香”，印度大吉岭红茶似麝香葡萄酒的芬芳，以及印度阿萨姆红茶浓烈的茶香，最能体会和欣赏它们的方式非清饮莫属。

Pure Tea Drinking: For a 100% Original Flavor

Pure drinking refers to the liquid tea with the original flavor 100% remained, free of any condiments, beverage or food. It has been the mainstream practice in China since Lu Yu advised in his momentous tea work the beauty of unmixedness. Kungfu Tea from Chaozhou, for example, is the best-known oolong drink for purity.

China is the homeland of tea and rich in germplasm resources of tea plant. The country's wide variety of tea-making techniques and people's obsession with refined tea ceremony should both take credit for pure drinking, the best way to access the truth of tea varying in species, technique and geo-conditions.

The same can be said of China's major six tea types and the premium black tea sorts from around the globe. Better way to cherish the sweet, orchid-like fragrance of Keemun, the musk wine smell of Darjeeling, or the thick flavor of Assam? Don't put anything else.

工夫茶，最为考究的清饮法

如若清饮，依茶类不同可选择冲泡、闷泡或煮饮，还有人用冰浸茶叶，或像冲咖啡一样冲滤茶叶。各种冲泡方法茶汤风味各异，其中最为讲究的，是潮汕地区的工夫茶，茶叶、茶具、准备、冲泡、品茶样样都蕴含“工夫”，茶汤也格外香醇。

潮州工夫茶是潮汕地区特有的饮茶习俗，有“中国古代茶文化活化石”之称，2018年被列入国家级非物质文化遗产名录。

潮汕工夫茶的流行始于明末清初。“工夫茶”不是茶叶名，而是泡茶品茶的技法。工夫茶茶具有特定的材质和样式，主要的四件茶具被称为工夫茶“四宝”，即孟臣罐（紫砂茶壶）、若深瓯（小的薄瓷茶杯）、玉书碨（烧水的陶壶，又叫砂铫）、潮汕炉（红泥或白泥炉）。潮汕工夫茶按照独特、考究的冲泡程式冲泡乌龙茶，人们常常说起的“关公巡城”“韩信点兵”就是潮汕工夫茶的程式。在潮汕人心中，茶事活动是一件很讲究、很认真的事情，精选的茶叶、特定的茶具、考究的冲泡、品饮程式以及礼仪，处处体现出“工夫”二字。

Kungfu Tea :A Flavor from a Sophisticated Process

For an authentic flavor, we would have to prepare a tea right - pouring, steeping or cooking, and some people even prefer to get it soaked in ice water or filter it like coffee. Of all brewing methods, Kungfu (Congou) Tea is particularly known for its sophisticated process. From what to how, everything in preparing a highly aromatic Kungfu Tea requires patience and dexterity.

Kungfu Tea, a custom unique to Chaoshan area, is known as a "living fossil of the ancient Chinese tea culture". It entered into the list of China's National Intangible Cultural Heritages in 2018.

Kungfu Tea became popular from the transitional period between the Ming and Qing Dynasties. Instead of a name of tea species, it actually refers to the techniques of tea brewing and sipping. The tea sets, usually made of special materials, are notably styled. The foremost four parts in a set are called "Four Treasures", namely the "*meng-chen-guan*" (purple-clay teapot), "*ruo-shen-ou*" (small, thin porcelain teacup), "*yu-shu-wei*" (pottery waterpot, also called "*sha-diao*") and "*chao-shan-hong-lu*" (red or white mud oven). To prepare a special oolong tea, Kungfu Tea observes a unique, sophisticated process named after some historical figure like "*guan-gong-xun-cheng*" (General Guan Yu on patrol) or "*han-xin-dian-bing*" (General HanXin in troop review). Brewing and serving tea seems particularly serious to Chaoshan locals. The name"kung-fu", literally "sophisticated process", is reflected anytime in the choice tea, select teaware, refined brewing and serving techniques and noble etiquettes.

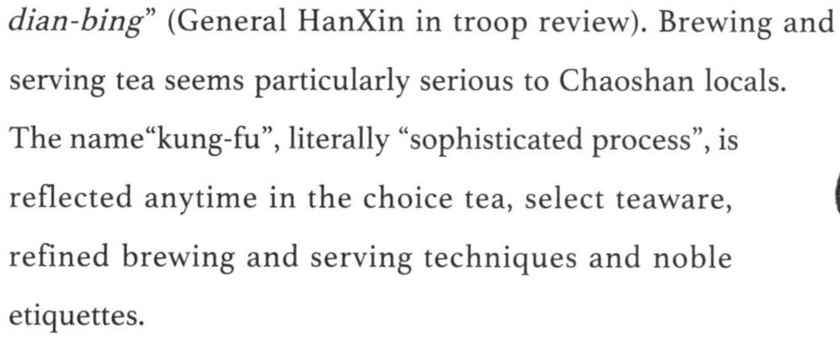

调饮茶，传统的花样融合

调饮茶与清饮茶相对，是以茶汤为原料，在茶汤中调入花果、奶类、糖或蜂蜜、香料等，根据时令及个人喜好灵活调配出的各种适口的茶饮。

相传，调饮茶源自文成公主远嫁吐蕃。高原的气候和肉类、乳品为主的饮食常令公主感到身体不适，于是她把带来的茶叶煮成茶水，加入奶再饮用，身体的不适得到缓解，茶叶与茶饮文化由此传入西藏地区。茶能解油腻、助消化、补充营养，藏族同胞将用茶汤、奶、酥油等调制的酥油茶视如生命，随时饮用。

中国的酥油茶、奶茶、擂茶、打油茶、三道茶、青豆茶、三炮台等，英国的奶茶，印度的马萨拉茶、甜奶茶，俄罗斯的甜红茶，摩洛哥的甜绿茶……调饮茶品种多样，极富特色，早已深度融入了人们的生活。

Flavored Tea: Age-Old but Diversified

Flavored tea, as opposed to pure tea, refers to a family of palatable beverages blending liquid tea with flowers, fruits, milk, sugar, honey or spices according to the season and one's palate.

Legend has it that flavored tea originated from Princess Wencheng, who married the Tibetan King Songtsen Gampo. Unaccustomed to the plateau climate and the local meat and dairy based diet, the princess cooked the tea she brought with her dowry and mixed the soup with some milk. Such "mixed tea" did work for the princess' health, and thus marked that tea and tea culture started to take root in Tibet. As tea can relieve greasiness, help digestion and supplement nutrition, Tibetans call buttered tea (made of tea, milk and butter) a priceless treasure, or a must-have in life.

Flavored tea comes in a good deal of sorts. Buttered tea, milk tea, "lei-cha", "da-you-cha", three-course tea ("san-dao-cha"), green-bean tea ("qing-dou-cha") and "san-pao-tai" are typically popular flavored teas in China, while milk tea (the UK), masala chai and sweetened milk tea (India), sweet black tea (Russia) and mint tea (Morocco) are also the world-renowned specialties, to name a few. In so many sorts, flavored tea has distinctively become a day-to-day existence for almost everyone today.

新茶饮，年轻人的创新百搭

新茶饮是为适应新时代年轻人天然、健康、时尚、方便的茶饮需求而出现并逐步发展起来的，其最大的特点是采用新鲜食材——茶汤萃取自茶原叶，食材使用新鲜的牛奶、水果、芝士、坚果等。新茶饮的品种主要有新式奶茶、水果茶、混合茶、纯茶、末茶等几大系列。

● 新式奶茶：以红茶、乌龙茶等搭配奶、珍珠粉圆、布丁、椰果等，现场加工成珍珠奶茶、炭烧奶茶、黑糖末茶牛乳、红茶拿铁，有的还在茶汤上覆盖“奶盖”，又好喝又好玩。

● 新式水果茶：以特殊香气的茶叶搭配特色新鲜水果，现场加工成色泽美丽，外观诱人的茶饮，如百香果茶、菠萝茶、雪梨茶等。

● 新式混合茶：以茶、花、果等搭配奶、芝士、可可、咖啡、酒等，现场调制成风味、外观别具新意的茶饮，如白桃乌龙、蔬果茶、果汁茶、鸡尾茶等。

● 新式纯茶：选用优质特色茶叶，用长时间冷泡或快速热泡等制作方式，现场加工出香气独特浓郁、滋味鲜醇可口的茶汤，大大提升了年轻人对纯茶的接受度。特色纯茶品种如纯绿茶、冷泡冻顶乌龙等。

● 新式末茶：以特色末茶搭配鲜奶或乳制品、糖浆、可可、冰块等，通过搅拌、打奶、打碎冰块等方法制成末茶拿铁、末茶星冰乐、末茶可可等。

In-Fashion Tea:
A Novelty for the Youngs

The so-called "in-fashion", like new-type milk tea, fruit tea, mixed tea, new-style pure tea and ground tea etc., refers to the use of fresh ingredients to make teas. Unlike old practice, now we use originally-flavored tea brew to match with fresh milk, fruits, cheese or nuts. They are coming on the scene to satisfy the growing population in quest for natural, healthy, stylish and ready-to-drink tea beverages.

Bubble tea, charcoal-roasted milk tea, brown sugar matcha milk and black tea latte are just a few of such "next-generation" milk teas that, based on black or oolong tea, are the mix with milk, tapioca pearls, pudding or coconuts to enhance flavors. Some of them are even made "milk (cream)-capped" as particularly a fun choice.

New-type fruit tea refers to a category of multi-colored, seductive beverages prepared as the mix of fragrant tea and fresh fruits, comprising passion fruit tea, pineapple tea and pear tea.

There is an in-vogue beverage mixing tea, flowers and fruits with milk, cheese,

cocoa, coffee or wine, extraordinary both in flavor and look. It comprises white peach oolong ("*bai-tao-wu-long*"), fruit & vegetable tea, juice tea and cocktail tea.

Even pure teas were updated. The new category, with premium tea either cold-brewed long or hot-brewed for a short while, features a distinctly rich smell and a mellow taste. It was a revolution to grow pure teas' popularity. Pure green tea and cold-brewed Dong-Ding Oolong, among others, are just a few specialties to name.

With ground tea as base, the "re-invented" category mixed with fresh milk, dairy products, syrup, cocoa or ice cubes by stirring, frothing or ice-breaking, comprises ground-tea latte, frappuccino, and ground-tea cocoa.

快捷饮，再忙也要喝杯茶

饮茶既是“琴棋书画诗酒茶”中的一桩雅事，也是“柴米油盐酱醋茶”中的寻常生活。茶可繁可简，闲暇时人们用心泡茶，细细品饮，泡茶与品茶的过程令人身心愉悦，忙碌中人们也需要一杯芳香的茶，抚慰身心、振奋精神。

为了让人们在忙碌中快速方便地喝到好喝的茶，20 世纪初，袋泡茶出现了，一个茶包、一杯热水，待茶色迅速在水中晕染开来，就可以喝了。近年来，杯泡茶进入人们的视野，在易降解的纸杯底部放茶叶，上置植物纤维滤纸隔离，冲入热水，能喝到茶水，又不用担心茶渣漂浮。还有一种速溶茶，是茶叶萃取、干燥制成的粉末，人们随身携带，想喝茶时可将茶粉倒入容器，用冷水或热水溶解，就是一杯香醇好喝的茶。

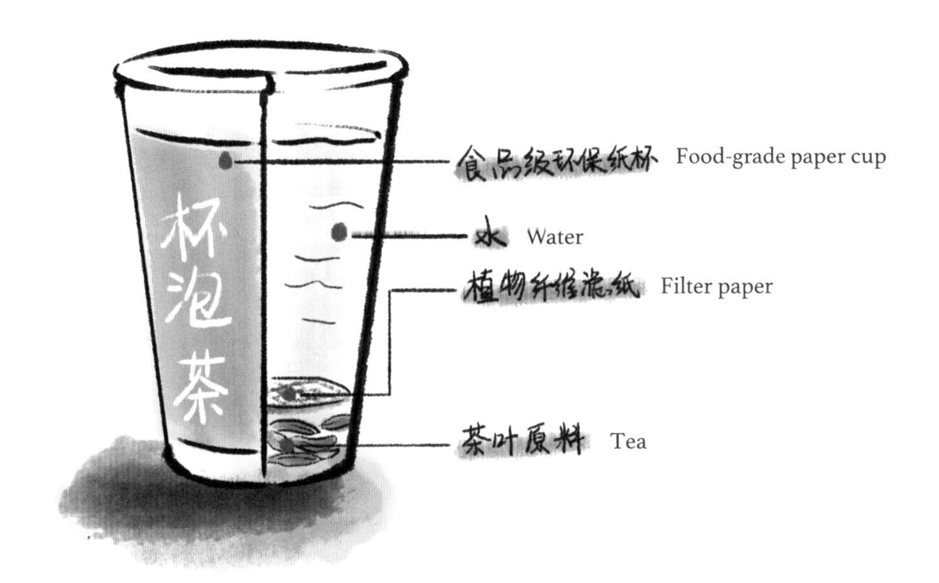

杯泡茶
Ready tea cup

Easy Tea: A Quickened Pace in Life

Tea can either be considered a graceful, well-born matter or simply be a cheap life stuff. Instead of the ceremonial complexity, we can also make tea quick and easy. Preparing a tea meticulously in free time seems like a big pleasure, while for much-occupied moments, in fact, we can't do without the magic beverage, either.

Tea bags appeared in the early 20th century as a gizmo to have saved us much time. Just soak the bag in hot water, and you would get the good-smelling liquid in a few minutes. "Ready cup" has been another hot cake over the recent years. Tea is stocked at the lower part and covered with a filter paper inside. Pour the hot water, and a fresh tea, free from dregs, is ready. The third invention is called instant tea, which is actually dried tea powder easy to carry. Use cold or hot water to dissolve it, and you would have a fragrant tea.

茶之形，美若浑然天成

茶叶种类丰富，外形更是多样。它们有的已经消失在历史的长河中，如唐宋的团饼形茶，有的被人们创制并流传。

历代的制茶工匠带着奇思妙想，做出形色如此美妙的茶叶——形似瓜子片的，如安徽的六安“瓜片”；形似眉毛的，如浙江“眉茶”、安徽“秀眉”、江西“珍眉”；形似小笋的，如浙江长兴“紫笋”；形状圆直如针的，如湖南岳阳“君山银针”、湖南安化“松针”；形曲如螺的，如江苏“碧螺春”；状如蟠龙的，如浙江临海“蟠毫”；形似竹叶的，如四川峨眉山“竹叶青”；犹如一朵朵兰花的，如安徽岳西“翠兰”。还有把一根根茶叶以丝线扎结成各种花朵形状的，如江西婺源“墨菊”、安徽黄山“绿牡丹”等。茶叶精工细制，颗颗却似浑然天成。

Tea Shapes: The Beauty in Miscellany

Species and variety aside, we also have a broad choice of tea shapes. Some have lost in shadows, like the cake-shaped teas prevalent in the Tang and Song dynasties, yet more of them have survived the past centuries in miscellaneous forms.

For such variegated shapes and colors, the ancestral wits would have to take credit. "Gua-Pian" from Lu'an (Anhui province), for example, are shaped like melon seeds; "Mei-Cha" from Zhejiang, "Xiu-Mei" from Anhui and "Zhen-Mei" from Jiangxi seem all like eyebrows; "Zi-Sun" from Changxing (Zhejiang province) is renowned for a strong likeness to bamboo shoots; "Jun-Shan-Yin-Zhen" from Yueyang and "Song-Zhen" from Anhua (both in Hunan province) are thin, straight like needles; "Bi-Luo-Chun" from Jiangsu looks spiral in shape; "Pan-Hao" from Linhai (Zhejiang province) resembles circling dragons; "Zhu-Ye-Qing" from Mt. E'mei (Sichuan province) takes after bamboo leaves; "Cui-Lan" from Yuexi (Anhui province) mirrors orchids in full bloom; and "Mo-Ju" from Wuyuan (Jiangxi province) and "Lv-Mu-Dan" from Huangshan (Anhui province) are shaped with silk threads into flowers..., to name just a few. Though being all carefully shaped, there seems no trail of hand-making to follow.

茶之器，
与茶天然契合的瓷搭档

泡茶需选用相宜的茶器，茶器不仅造型多样，材质也极为丰富，如陶、瓷、竹木、玻璃、金属等，其中瓷茶器最具代表性。瓷器是中国伟大发明之一，与茶叶同为中国带给世界的重要礼物，自唐代起，两者就相伴走向世界。用瓷茶器泡茶品茶是人们最自然的共同选择。

中国人认为瓷器汇集了天地五行与人的智慧，因而对它有天然的喜爱和亲近之感。茶与瓷天生为友，它们都带有质朴天然本色，都生自泥土，蕴含大自然精华，经历高温的洗礼，成于匠人之手。从古至今，瓷茶器是茶器中的绝对主力，每个时代都留下了经典瓷茶器的传奇，如唐代越窑、邢窑等烧制的茶碗、执壶；宋代汝窑、定窑、官窑、哥窑、钧窑五大名窑烧制的茶盏、盏托、汤瓶等，以及建窑的建盏等黑釉茶盏；明清以来，景德镇窑等瓷窑烧制的仿古、创新瓷茶器品种花色之丰富令人爱不释手。瓷茶器光洁闪亮、润泽如玉、历久弥新，为全世界所钟爱。

Porcelain: A Gift Called "China" from China

To make a good tea, teaware does matter. Assorted in size and shape, teaware can either be made of ceramics, porcelain, bamboo, wood, glass or even metal. Porcelain is no doubt the most quintessential one. Porcelain is a historic discovery and, along with tea, a gift China offered to the entire globe. Tea and porcelain came on the scene as a pair from way back to the Tang Dynasty, thus making the latter naturally the first choice across the global village.

Revered as an epitome of the occult "Five Elements" and human wits, porcelain has always been a national treasure in China. Tea and porcelain are true friends. They are both no-frills creations from the soil, or the essentials of the nature. Both have to weather high temperatures before taking shape in the most skilled hands. Of all periods, porcelain has played the cornerstone in tea sets, and legends of porcelain teaware were not missing for each phase. Tea bowls and kettles ("*shui-zhu*")from Yueyao and Xingyao kilns, for instance, were in vogue across the Tang Dynasty. Back to Song China, the "Five Kilns", i.e, Ruyao, Dingyao, Guanyao, Geyao and Junyao, were renowned for making tea cups, saucers and pots. Black-glazed utensils, like "Jianzhan" from the Jian Kiln, were also an option in the same period. Jingdezhen Kiln began to take center stage from the Ming and Qing Dynasties as its archaized and novel porcelains, so rich in style and pattern, were being mass manufactured. Porcelain teaware has become a global pick for its invulnerable, glossy traits.

茶之技，怎么泡茶才好喝

如何把茶泡得更好喝，这是一个需要反复尝试、摸索和总结才能回答的问题。一杯好喝的茶汤离不开茶、水、火、器、境、人等的通力协作。

茶叶的品质是一杯好茶的基石，不同的茶类和不同品质的茶叶决定了适合它们的冲泡方式。

水为茶之母，好水具有“清、轻、甘、冽”的特性，能提升茶汤的滋味。

“活水还须活火烹”，古人非常注重用火的精妙，他们观察煮水时随着水逐渐沸腾，从锅底冒出的水泡从小如虾目，至大如蟹目、鱼目直至水沸腾如鼓浪，借此把控烹茶水温。现代的科技手段帮助我们能够更直观地了解和控制水温，从而保证茶汤的香气和滋味。

器为茶之父，选用适合的茶器能使茶汤更香醇。

饮茶环境对饮茶者心境有着潜移默化的影响。同样饮一杯茶，在春天落花的桃树下，在夏季清凉的书斋里，或是在中秋月圆之夜、冬日飘雪的林间，会带给人不一样的心境。

此外，泡茶人的沏茶技巧，对水温、投茶量、泡茶时间等的把控，直接决定了一杯茶汤的风味和带给人的感受。

Tea Knowledge: How a Fine Tea is Prepared

How to make a perfect tea is a question that merits repeated attempts and review. A fine liquid tea usually culminates from the concerted play of dry tea, water, fire, utensils, environment and, of course, the person preparing it.

The dry leaves lay the cornerstone for the brew quality, while the brewing methods usually rest with the tea sorts and their varied traits.

As the "mother" of tea, "clear, light, sweet and pure" water helps thicken the fine taste.

"Great water should be on wise fire." In old days, the subtlety of fire control requests fine-tuning the water temperature by closely observing the air bubbles, growing from shrimp eyes, crab eyes, fish eyes to raging torrents. Today, however, we use modern instruments to read the temperature and put it under control. It's proved helpful to retain the soup fragrance and taste.

Teaware is the "father" of tea, as a felicitous tea set can augment the aroma.

Environment also exerts a subtle influence on one's mood. A sip would mean quite different for someone under separate circumstances, say, sitting by a full-flowered peach tree in spring, at a refreshingly cool study room in hot summer, over a moon-lit night, or amidst a snowy wood in winter.

Skills like the control over the temperature, dry tea intake and brewing time could also affect the tea flavor and even one's mindset.

茶之益，爱喝茶的人更健康

茶不是药，但喝茶有益健康则是不争的事实。陆羽的《茶经》中收录了多条古人关于茶饮健康的历史资料，认为茶有解毒、令头脑清醒、帮助消化等作用。

现代科学研究发现，人体中的自由基积累过多会对机体造成伤害，故有“自由基是万病之源”的说法。过多的自由基会引发一系列疾病，如癌症、心脑血管疾病、炎症、白内障、糖尿病、老年痴呆等。而茶叶中的茶多酚清除自由基的活性较为强大，因此茶多酚也被认为是茶中最主要、最精华、对人体最有用的成分，饮茶对健康的益处主要体现于茶多酚的功效上。此外，茶叶中的咖啡因能消除疲劳，使人头脑清醒，同时能刺激胃酸分泌，帮助消化、降脂轻身；茶氨酸有镇静安神的作用，可以平和身心。

Tea Benefits: A Cure for Health Problems

Tea is no medicine, but it is indisputably good for health. In "*The Classic of Tea*", Lu Yu enumerated historical accounts or anecdotes about the benefits of tea, holding that tea could help detoxify body, sober mind and boost digestion.

Scientists have warned that excessive accumulation of free radicals is harmful to our health, hence a saying goes that "free radicals are the root of all diseases". An excess of free radicals will invite a series of diseases, such as cancer, cardiovascular disease, inflammation, cataract, diabetes and senile dementia. Endowed with high radical-scavenging activity, tea polyphenols found in tea leaves are the most beneficial component for humans. Tea drinking thus means somehow to stay fit by killing free radicals. Besides, caffeine in tea can help dispel fatigue, awaken mind and stimulate gastric acid secretion for digestion, fat-burning and weight loss. Theanine, another tea chemical, can help steady nerves and is therefore used to calm one's mind and body.

茶空间，唯美的身心休憩场所

茶空间是根据茶文化内涵设计、布置的专门用于品茶的场所。人们在茶空间里以茶会友，陶冶身心，感受茶文化的氛围与美感。有人、有茶、有茶器，就有了饮茶的空间。古代文人为了品茶论道、修身养性建造茶室、茶寮、茶亭。他们的理想是依山傍水构斗室一间，室外茂林修竹，室内净几明窗，内设茶具，焚香、抚琴、品茗、看画……这种人与自然和谐共处，将茶饮与生活、与文化相融的追求，至今仍对人们的茶事生活产生极大的影响。

如今，茶空间不拘地点，庭院、茶馆、书房，或山间水边、松下林旁，人们或有雅器、书画相伴，或被大自然拥抱，有良友，有好茶，茶空间成为令人身体舒适、心灵得以休憩的独特场所。

Tea Space: A Chamber Close to Mother Nature

Tea space is a location deliberately furnished in quest for the ultimate understanding of tea. It takes in tea parties, refining people's mind with an aura of whatever makes tea so profound. This is properly a mix of tea, tea sets and those who love tea. In ancient China, tea huts, houses or pavilions were knowingly the "spaces" to pursue a state of spiritual entirety. It could be ideally a cubbyhole nestling into the picturesque hills and creeks, within an immediate reach of lush woods outside and plainly outfitted indoors, for a life simply with incenses, music, tea and paintings. Such agreement with the nature and the unity in life and culture still have a significant impact on our thought of tea today.

Now a tea space is not confined to anywhere. It could be a courtyard, a teahouse or evenone's daintily-adorned study room, or just an opening neighboring waters or thick woods into the arms of the nature. With soothing teas and bosom friends, the space will be a unique setting for physical and psychological peace.

茶旅行，探访茶的前世今生

茶被世人喜爱，吸引人去寻访它的故乡和足迹，由此形成了以茶文化为主题的寻茶之旅，并迅速在爱茶人中发酵。

跟着茶叶去旅行，愉悦身心、体验文化、感受自然与人文之美。人们奔赴茶山，走入茶园，走访与茶相关的人文古迹，感受茶叶的来之不易和茶的丰富历史文化内涵。

来到茶叶生长的灵秀之地，人们亲手采摘茶叶，看茶农劳作、体验做茶，参观主题庄园和博物馆，了解产茶地人文茶俗……随着深入茶乡，人们对茶的好奇与探究渐渐转变为对大自然造物的感叹、对古今制茶匠人的赞美，并与古往今来的爱茶人产生共情。

Tea Visit: A Journey for Tea Understanding

Tea, as universally cherished, attracts numerous visits to its home left with intangible remains. A tour in search of tea's cultural existence has thus been gaining popularity, particularly among tea fanatics.

A refreshing journey spiced up with teas helps unveil the spectacular nature and human culture. Exposure to tracts of tea farms, with a search of sites with varied tea legacies, invokes an understanding of hardship behind tea fragranceand the richness of intangible tea heritages.

The route may end up tea gardens to pick leaves, watch and try the processing, visit the tea village and museum, and access the local tea science and customs... Deep into the farms, curiosity and inquiries start to give way to the marvels of creation, admiration of all fine tea makers in history, and an emotional state shared with tea lovers of all ages.

茶社交，一杯茶融合彼此

从三国时（220—280 年）起，人们就喜欢聚在一起喝茶言欢，古人称之为茶宴或茶会。至唐代，宫廷、官吏、僧侣、士人及百姓的各种茶会雅集蔚然成风。《宫乐图》描绘了一场优雅的唐代宫廷女子的茶宴场景；宋徽宗的《文会图》描绘了宋代文人雅士品茗雅集的场景，他曾亲自点茶宴请群臣；清乾隆皇帝举办过四十余次重华宫茶宴，并独创“三清茶”……与宫廷茶会典雅、庄重不同，民间茶会则更加清雅、平实，茶会上人们喝茶、享用茶点，有时还穿插赋诗、焚香、抚琴等文化活动；有人则喜欢去茶馆聚会，一壶茶，畅谈古今中外。起源于唐朝中期，兴于宋元时期的径山茶宴流传至日本，成为日本茶道之源。

当茶文化走向世界，茶的亲和力得以进一步体现。17 世纪兴起的英国下午茶很快成为当时人们的重要社交活动，大家一起喝茶、吃传统的英式点心，交流各种见闻，谈天说地，享受下午茶的美好时光。

时至今日，几杯清茶，几盘茶点，新朋或老友围坐一桌，聊聊茶，说说工作和生活，心的距离迅速靠近。以茶为礼、以茶待客、以茶会友，茶是融洽彼此不可替代的琼浆玉露。

Tea Party: From Privacy to Social Connections

Tea party, or tea banquet, first appeared in China around the Three Kingdoms Period (AD 220-280), and it became a common exercise for people from all classes in the Tang Dynasty. It prevailed even for imperial families: the scroll painting "*The Palace Music*" dating from Tang China depicts a women's tea party in the court; Emperor Huizong, in his masterwork "*The Function*" , also reproduced the tea parties in which he even made teas in person for his ministers; and Emperor Qianlong held tea banquets over 40 times at Chonghua Palace and created by himself "Sanqing Tea"... to name a few. Not that necessarily dignified, the party for average classes were somehow restrained. Fine tea and tea snacks were served, and sometimes with a request for impromptu poems, incense burning or gu-qin play. Teahouse became also a place for regular meetings and chitchats. The renowned Jingshan Tea Feast, which first appeared in the middle of the Tang Dynasty and gained popularity in Song and Yuan Dynasties, was introduced to Japan and thus became the prototype for the sophisticated Japanese tea ceremony.

Tea became essentially a catalyst for global social connectionseither as it was introduced to the world. The British afternoon tea since the 17th century has become an inextricable part in people's regular networking. Men and women consumed teas and refreshments, while usually in the exchange of news or gossips, to spend a good time with families or friends.

A tea with a few snacks would be enough for hours of talks, either about work or life. A morphed existence with social personality, tea plays as a gift, a treat oran excuse to meet friends. In short, tea is the answer to much-valued, wide social connections.

Chapter Four

—

The Spread of Tea
A Rich Smell to Please the Entire World

第四部分

茶的传播

和合世界香飘远

引文

茶聚四海知己，缘和天下亲朋。中国是茶的故乡，茶从中国出发，传遍五大洲。以中国为起点，通往世界各地的多条茶叶之路上流淌着甘美的茶水，也交流和传递着缤纷多彩的茶饮文化，各国独特的茶饮习俗犹如颗颗珍珠，熠熠发光。

茶带给世界的，不仅是繁荣的经济、美好的生活，更有各地独特的茶文化。中国茶文化在日本演变成日本茶道，深刻影响着日本人的精神世界与衣食住行；同为中国近邻，韩国的茶礼秉承中国儒家礼制的思想并影响着韩国人的生活习俗；在英国，下午茶充满西方式的惬意优雅；俄罗斯人习惯以茶炊煮饮浓浓的红茶；摩洛哥人则用长嘴巴、大肚子的茶壶冲泡薄荷甜绿茶……茶成为这些国家独具特色的文化符号，它是中国给全世界最诚挚的友情卡。

Introduction

Tea makes people friends. China is the hometown of tea, and gradually tea was spread all over the globe. All starting from China, the prolific trans-border tea routes were shaped into a sprawling web for both worldwide trading and cultural exchanges, giving rise to a broad spectrum of tea customs that each sets apart from the others.

To the entire globe, tea not only brings economic and life prosperity but inspires unique tea cultures from separate places. In Japan, the tea ceremony originating from the Chinese tea culture has reshaped the Japanese' life and soul in all respects, while in South Korea, another neighbor of China, the tea rite reflecting the Confucian thoughts has remained a part of Korean living. A classic afternoon tea is considered by the British a blue-blooded code.In Russia, superbly strong black tea prepared with a samovar is most preferred, while people in Morocco love to make what is named Maghrebi Mint Tea with a big-bellied pot. For all those magnificent cultural hallmarks, tea is a blessing from China, or plays as China's ambassador of friendship, to the world.

金大廉公茶始培追远碑（左侧石碑）

The stele in memory of Kim Taeryom, the one first brought back tea seeds to the Korean Peninsula (left)

茶籽传入朝鲜半岛，1000 多年前的御赐礼物

自古以来，中国与接壤的朝鲜半岛在政治、经济、文化等方面交流较为密切。善德女王时期（632—647 年），中国的茶叶和茶文化随佛教传入朝鲜半岛。中国盛唐的茶文化兴盛时期，大量来自朝鲜半岛的僧人、学者到大唐修学，学成后，他们将茶籽、饮茶器具及饮茶方式带回朝鲜半岛。

朝鲜半岛史籍《三国史记》中记载，兴德王三年（828 年），从大唐返回的大使金大廉把皇帝御赐的茶籽带回朝鲜半岛，兴德王命人将茶籽种在地理山下。至高丽时期（918—1392 年），朝鲜半岛的饮茶习俗普及到社会各个阶层，茶产业与茶文化均发展繁盛。如今，智异山（又名地理山）立有金大廉带回茶种开始种茶的纪念碑。

A Gift for the Korean Peninsula More Than 1,000 Years Ago

China has had close bond with the bordering Korean Peninsula in politics, economy and culture. Tea and tea culture were first introduced, along with Buddhism, from China to the peninsula during the reign of Queen Seondeok (632-647). As the Tang Dynasty came to its heyday, the country's tea culture was in blossom. Numerous monks and scholars from the peninsula went to Tang China for study. When mission completed, they returned home with tea seeds, utensils and even the brewing methods.

According to Samguk Sagi (*History of the Three Kingdoms*), Kim Taeryom, the King Heungdeok's envoy to the Tang Dynasty, brought tea seeds from the Chinese emperor back to the peninsula in AD 828. The King had his men plant them at the foot of Mount Jiri (or Jirisan). As tea became universally accepted during the Goryeo Period (918-1392), tea-related economy and culture were growing fast then. We may still find the monument in honor of the envoy's groundbreaking contributions, in Jirisan today.

茶文化传入日本，805年起，日本僧人带回茶籽、茶具与饮茶法

中国茶和茶文化由来中国学习的日本僧人传入日本。据记载，630年到894年，日本赴大唐的遣唐僧将茶叶带回日本。自805年起，最澄、空海、永忠等遣唐僧将中国茶籽、茶具、茶典和饮茶方式等带回日本进行推广。他们将茶籽试种于日本滋贺县日吉神社，现在京都比睿山东麓立有“日吉神社御茶园碑”。

12世纪，中国茶文化再次被引入日本。1168年、1187年，荣西禅师两次到大宋天台学习禅宗，回国后将从中国带回的茶籽种植在背振山并栽种成功，茶树在日本开始广泛种植。荣西还撰写了日本第一部茶书《吃茶养生记》，被誉为“日本的《茶经》”。

1241年，僧人圆尔辨圆从浙江径山带回《禅院清规》、径山茶种和饮茶方法。1259年，南浦绍明将径山茶宴系统地带回日本。此后，历经村田珠光、武野绍鸥和千利休等人的完善，具有日本本土特色的茶道逐步形成。

The Introduction of Tea Culture to Japan Since AD805: Seeds, Utensils and Customs

Tea and teaculture were introduced to Japan by student-monks the Japanese government sent to China. Tea went first to Japan over a period from 630 to 894, as said in documents. The monks, since AD 805, brought tea seeds, teaware, tea book and tea customs back to the Japanese people, and Saicho, Kukai and eichyuu (youchyuu) were the three renowned of them. They tried to grow tea at Hiyoshi Shrine in today's Shiga Prefecture. The remains of Hiyoshi Imperial Tea Garden, together with a stele, still exist at the eastern foot of Mount Hiei, Kyoto.

The Chinese tea culture was landed to Japan again in the 12th century. For Zen learnings, Japanese monk Eisai went to Mount Tiantai twice in 1168 and 1187. Back to Japan, Eisai scattered the seeds he took from China in Mount Sefuri. He succeeded, and henceforth tea groves were spread over the country. Kissa Yojoki (*Notes of Tea for Health*) was another legacy he left. It was Japan's first tea book, or known as the "*The Classic of Tea*" of Japan.

Enni Ben'en was a visiting monk in China. In 1241, Ben'en returned to Japan with the Regulations in *Zen Learnings*, tea seeds and brewing practice he deliberately collected from Jingshan in China's Zhejiang province.In 1259, monk Nanpo Jomyo introduced the entire "Jingshan Tea Feast" back to Japan. The rite was henceforward perfected by Murata Juko, Takeno Jo'o and Sen Rikyu before "Sado", the tea ceremony with Japanese characteristics, finally took shape.

茶传欧洲，
1610年，荷兰商船运茶回到欧洲

1607年，荷兰商船自爪哇到澳门运载茶叶，1610年，由澳门运茶到欧洲，打开了中国茶叶销往西方的大门。不久，茶叶成为荷兰最时髦的饮料，饮茶之风也迅速风靡英、法等国。自此，茶叶不断从海路输入欧洲，饮茶之风从荷兰兴起，逐渐传遍整个欧洲。

首先带动欧洲茶饮潮流的是英国王室。1662年，嗜好饮茶的葡萄牙公主凯瑟琳嫁给英国国王查理二世，将中国红茶带入英国宫廷和上流社会，并且带动了英国全社会的饮茶之风。18世纪，英国安妮女王常邀请贵族共赴茶会，还请人特别订制茶具、瓷器柜、小型活动桌椅等，形成了优雅素美的“安妮女王式”饮茶风格，从而推动了英式下午茶的流行。茶的传入为欧洲人民的日常饮食和休闲生活提供了又一美好的选择。

The First Landing in Europe in 1610

Dutch mariners, from Java, came to Macau first in 1607 for tea business. The shipment arrived in Europe in 1610, thus opening the doors to the west. Tea shortly became a phenomenal drink in the Netherlands, and later in vogue across the British Isles and France. The maritime route was made an inexhaustible source of tea for Europe. It became widely embraced first in the Netherlands and later all over the continent.

It was the British royal family that played the role model. Catherine of Braganza, the daughter of King John IV of Portugal, married King Charles II in 1662. The black tea from China she brought inspired the tea-drinking fashion in the British court and among people of higher social standings, and later across the whole Britain. In the early 18th century, Queen Anne was a regular hostess of tea parties for her ministers and lords. Creating the vintage "Queen Anne Style" by using tailor-made tea sets, china cabinets and portable tables and chairs, she was widely reckoned in the vanguard of the British afternoon tea. For Europeans, tea became the best option either on dining tables or as an escape from their jobs.

茶传美洲，
1690年，波士顿有了中国茶叶市场

17世纪中期，茶叶随同欧洲移民传到美洲大陆，很快，茶就成为当地人最常见的饮品。1690年，波士顿就有了销售中国茶叶的市场。18世纪20年代，北美洲殖民地开始大量进口茶叶，此后饮茶风气陆续传遍美洲各地。据统计，到18世纪60年代，中国茶叶在北美洲每年销售量约为120万磅（1磅约为454克）。

一片小小的叶子也能掀起巨大的波澜。1773年12月16日，为反抗英国国会颁布的茶税法，北美洲爆发了历史上著名的“波士顿倾茶事件”。波士顿倾茶事件在一定程度上推动了美国革命的爆发，最终改变了世界。

The Chinese Tea Market First Seen in Boston in 1690

Tea made its first landfall on America with European immigrants in the mid-17th century. It seemed a sensational beverage overnight. Boston opened its first Chinese tea market in 1690. As massive tea imports to the colonies of North America startedin the 1720s, tea was spread all over the continents. For the Chinese tea in North America, the sales volume could reach 1.2 million pounds annually in the 1760s.

The butterfly effect also works for tea. To protest the Tea Act of Great Britain, the "Boston Tea Party" broke out on December 16, 1773. As tension was mounting, the incident became somewhat a catalyst for the outbreak of the American Revolution. Tea did play a role in such far-reaching events that changed the world.

万里茶道，17 世纪，茶叶从崇安走向俄罗斯

17 世纪至 20 世纪初，中国和俄罗斯之间曾有过一条“万里茶道”。这是一条多起点、多终点，在不同时期先后兴起，呈网状分布的人文与商贸通道，是中俄等国文化交流的载体和桥梁，也是东西方文化交流、文明互鉴的纽带。

“万里茶道”是连通中俄两国的“世纪动脉”。它的起点在福建崇安（现福建武夷山市），茶叶最终销往俄罗斯的西伯利亚、莫斯科、圣彼得堡等地，以及其他欧洲国家，全程两万余里，行程两年。万里茶道上运输的主要是福建出产的红茶和湖南、江西、湖北等地出产的黑茶。现在的蒙古国、俄罗斯等国茶饮习俗的形成与万里茶道有着密不可分的关系。

A Long Trek from China to Russia in the 1600s

The "China-Russia Tea Road", alive from the 1600s to the early 1900s, was a trans-border passage for both cultural exchanges and commercial use. It had several points to start and ended up separate cities. The sprawling web of the routes, each in the primeat separate times, was a bridge between China and Russia and abond between the eastern and western civilizations.

The Road was also an economic artery between the two powers. From Chong'an (today's Wuyishan, Fujian province), it would be a long tough journey finally to Siberia, Moscow, St. Petersburg and other places in Russia as well as other European countries around. The trek spanning over10,000 kilometers would cost two years. Black tea from Fujian and dark tea from Hunan, Jiangxi and Hubei were the major exports along the route. For today's tea popularity in Mongolia and Russia, the Tea Road should take credit.

茶传非洲，重要的茶叶产销地

20 世纪初，茶树引种至非洲的肯尼亚、坦桑尼亚及津巴布韦等国，并逐步大面积种植。此后，非洲茶业迅猛发展，与亚洲共同成为世界茶叶的重要出产地，肯尼亚、马拉维、乌干达、坦桑尼亚和莫桑比克等十余国为非洲重要的产茶国。

非洲东部的肯尼亚现在已成为全球产量第三、出口量第一的茶叶生产大国。茶产业是肯尼亚的支柱产业，为国家的发展提供着蓬勃的动力。肯尼亚有 500 万人从事茶叶生产，人们通过生产茶叶改善了生活。

除生产与出口大量茶叶外，埃及和摩洛哥的茶叶进口量分别位居全球第四位和第七位，两国虽然不产茶叶，但人们酷爱饮茶。摩洛哥人独爱饮中国的绿茶，是进口中国茶叶最多的国家。

Africa: A Robust Engine for the Global Tea Industry

Camellia sinensis, or tea plants, were introduced to Kenya, Tanzania and Zimbabwe in the early 20th century. Tea farms began to grow in size, henceforth seeing a boom of tea industry in Africa. The continent, together with Asia, has become the world's major tea supplier. Kenya, Malawi, Uganda, Tanzania, Mozambique and several other countries are nowadays the African countries for tea production.

Kenya in eastern Africa is the world's third largest tea producer and enjoys the largest tea exports. Tea is the backbone and power source of the country's economy. In Kenya, the population devoted to tea production reaches five million, making tea a business that has saved numerous Kenyans from poor living conditions.

Africa is more than a tea supplier. On worldwide tea imports, Egypt ranks the fourth and Morocco the seventh. They are not tea growing countries, but their people are desperate for it. Since people of Morocco are particularly fond of the Chinese green tea, the country has become the foremost tea buyer to China.

茶马古道，千年川流不息的国际贸易通道

今天人们常说的“茶马古道”源自唐代的“茶马互市”，兴于唐宋，盛于明清。它连接了中国产茶区和整个大西南地区，向境外延伸至东南亚、南亚和中亚地区，以马帮、背夫将茶叶从四川、云南等茶产地运往遥远的目的地，在古代中国对外经济文化交流和古文明传播中起过重要的历史作用，曾是中国历史上最为著名的西部国际贸易古通道之一。

虽然随着时代的变迁，这条历史上的重要商贸通道上已繁华不再，但茶马古道仍被视为世界上自然风光最为奇丽壮阔、人文景观最为丰富多元的神秘之路。

A Passage Once Thriving on Horseback

The name "Chama Gudao (Ancient Tea-Horse Road)" today originated from the fairs trading tea and horses. The fairs first appeared in the Tang and Song Dynasties and came to its heyday during the Ming and Qing. Connecting China's tea-growing areas to the vast southwest China and even to Southeast Asia, South Asia and Central Asia, the Road saw numerous caravans traveling on foot or on horseback, delivering tea from Sichuan or Yunnan to distant places. Being historically an exalted route for cross-border trade in western China, the Tea-Horse Road played a big role in the country's economic outreach and cultural networking.

The magnificent trade passage, as time flies, has gone in obscurity. However, it is still considered a road of mystery along with the world's most spectacular natural wonders and cultural legacies.

日本茶俗，茶道规范社会行为

唐宋时期，来中国学习的日本僧人将中国径山禅宗寺院径山寺的茶宴仪轨带回日本，将日常茶饭与宗教、哲学、伦理和艺术相融合，逐步形成了本土特色的日本茶道。千利休（1522—1591年）是日本茶道的集大成者，他提出了“和、敬、清、寂”的茶道精神。

日本茶道分为抹茶道和煎茶道两种，以抹茶道为主流。日本茶道具有严格、复杂的程式，一般在茶室中进行。茶师按规定动作点炭火、煮开水、冲茶或点茶，然后依次给宾客献茶，客人按规定礼仪品茶。按照习惯，品茶后客人要对茶具进行鉴赏和赞美。茶道是日本民族文化的代表，日本人把茶道视为一种修身养性、提高文化素养和进行社交的手段，它对日本人的精神文化、哲学、艺术、社会行为直至日常生活都产生了深远的影响。

Japan: The Special "Way of Tea" as Social Etiquettes

It was the visiting Japanese monks for disciplinary learnings who brought back to Japan the Jingshan Buddhist Temple's "tea feast" during the Tang and Song Dynasties. The Japanese Sado (tea ceremony) was the result of their fruitful thinking about tea with religious doctrines, philosophies, morals and arts. Sen Rikyu (1522-1591), a Sado master, first introduced the four virtues, namely "inner harmony, respect, purity and tranquility", behind the Sado practice.

Sado can be divided into matcha and sencha ceremonies, and in Japan the former prevails. Strict and complicated in procedure, Sado is usually staged in a tea room. For the host, there are established rules to set charcoal alight, boil water and steep or grind tea. As the made matcha brews are served in turn, the guests should follow some etiquettes before taking a sip. Afterwards, the ritual also requests their admiration for the host's tea utensils. Representative of the Japanese culture, Sado stands for a chance of self-cultivation, education and social meetings. It has posed a profound impact on the Japanese' core spirit, philosophy, arts, behaviors and even life routines.

韩国茶俗，各种茶礼以及各种“茶”

新罗时代（668—892 年），居住在朝鲜半岛的人们效仿中国佛教茶礼和规范，开始形成和建立了具有本土特色的古代茶礼。后来，被奉为“韩国茶圣”的草衣禅师（1786—1866 年）吸收中国儒家思想，倡导“中正”精神，为现代韩国茶礼“和、敬、俭、美”的精神奠定了基础。韩国茶礼种类多且各具特色，按茶叶类型分有“抹茶法”“饼茶法”“煎茶法”“叶茶法”等。每年 5 月 25 日是韩国“茶日”，要举行成人茶礼和高丽五行茶礼等活动。

韩国传统“茶”饮种类繁多，可谓无物不能入茶。常见的有五谷茶、药草茶、水果茶等，如大麦茶、玉米茶、艾草茶、葛根茶、大枣茶、核桃茶、莲藕茶等。其中以大麦茶最为出名——烘烤过的大麦粒放在开水中泡制，香气诱人，能去油解腻，健脾胃、助消化。

South Korea: The Assorted Rites and Teas

The ancient Korean tea rite, by referring to the Buddhist tea ceremony and its rules and etiquettes in China, took shape during the Unified Silla period (668-892 A.D). In South Korea, Master Choui(1786-1866) is known as the "tea sage". He borrowed the Confucianism with his own belief in "harmony", which was the beginning of the modern "tea rite" that emphasizes "harmony, respect, frugality and beauty". The rite comes in assorted names and characteristics. By category, we call it the rite of powder tea, cake tea, cooked tea or leaf tea. In South Korea, May 25 is a day for the hosting of "the Rite of Passage" and "the Rite of the Five Tea Elements".

The Korean "tea beverage" also comes in a wealth of categories and sorts. Almost everything looks fit for tea. Koreans people usually make grain tea, herbal tea and fruit tea, and being particularly notable are barley tea, corn tea, wormwood tea, radix puerariae tea, jujube tea, walnut tea, lotus root tea, etc. Exceptionally noted is the barley tea. Giving an inviting smell, the brew of roasted barley grains helps digest greasy food and maintain a good appetite.

土耳其茶俗，子母壶与郁金香杯

土耳其被称为“一个被浸泡在茶水中的国家”，土耳其人无论居家、外出都离不开茶，甚至在商店里，茶都是必备的饮品。人们爱喝加糖不加奶的红茶。煮茶工具很特别，为一大一小的“子母壶”，大壶煮水，小壶煮茶。饮茶的杯子也很特殊，玻璃材质，形状似身材窈窕的女郎，俗称郁金香杯。

在土耳其街头随处可见茶郎们或肩背或手提一个精致的金属托盘，盘子里有装着热茶的茶壶、郁金香杯、小托盘和方糖，这种特别的卖茶方式叫“走卖”。土耳其语中“茶”的发音和中文发音一模一样，所以闹市上时时会听到“Cha、Cha”的叫卖声。

Turkey: The Stacked Teapot and the Tulip-Shaped Glass

Turkey is literally a country "living on tea". Life in Turkey, either at home or not, is inseparable from tea. The beverage is even served in every shop. Turks love black tea with sugar, but they don't mix tea and milk. In Turkey, two kettle pots are stacked together for stovetop brewing. The big one is for heating water, the small one is for tea cooking. Tea is usually offered in glasses that look like girls in slim-thick physique. We usually call them "tulip-shaped" glasses.

Street cry of Turkish black tea can be heard everywhere. Hawkers travel around with a handsome metal tray on shoulders or hands, containinga filled teapot, tulip-shaped glasses, saucers and sugar cubes. This is called "Walk& Sell". In Turkish, tea is pronounced "çay". This is exactly the same as "cha", what tea is called in Chinese. When hearing the cry of "cha, cha" somewhere in Turkey, you would never miss a cup of sugared black tea.

印度茶俗，甜奶、马萨拉茶与拉茶

180 多年前，中国茶引种至印度，开启了印度的种茶、饮茶历史。如今，印度已成为产茶大国和茶叶消费大国。红茶是印度人的最爱。他们喜欢在红茶中加入奶制品和砂糖制成“甜奶茶”。有的人喜欢喝在红茶中加入姜、豆蔻、茴香、丁香、肉桂等的混合辛香料马萨拉的“马萨拉茶”，也叫“印度香料茶”。

为使马萨拉茶中的茶、奶与马萨拉的味道完美融合、泡沫丰富，口感细腻有层次，印度人将调饮茶从一个金属容器倒入另一个金属容器，循环往复地“拉”，这就是印度人的心头爱——拉茶，拉茶的制作在印度随处可见，可谓街头一景。

India: The Milk Tea, Masala Chai and Teh Tarik

The Chinese tea was brought to India some 180 years ago, thus ushering in the history of tea growing and drinking there. Now India is the world's major tea producer and consumer. For Indians, black tea is their favorite. They prefer mixing it with dairy foods and sugar. "Masala Chai" is another beloved beverage nationwide. Masala is a spice mixture in black tea or other Indian foods, usually comprising ginger, cardamom, fennel, cloves and cinnamon. It is also called "Indian spice tea".

To make a perfect blending of tea, milk and Masala, Indians invented Teh Tarik. Literally meaning the "pulled tea", this is made by cooling the mixed brew through the process of pouring and "pulling" it between two metal cups or mugs to create a rich, frothy drink. In India, bazaars and streets abuzz with Teh Tarik vendors are usually the backdrop to millions of tourist photographs.

斯里兰卡茶俗，酷爱清饮浓红茶

过去，斯里兰卡被称为“锡兰”。锡兰高地红茶与中国祁门红茶、印度的阿萨姆红茶、大吉岭红茶并称世界四大红茶。自 1824 年从中国引种茶叶至今近 200 年间，茶业已成为斯里兰卡农业支柱产业，茶叶产量、出口量均名列世界前茅。

茶是斯里兰卡人生活的必需品。在城市和农村随处可见茶站，人们付钱后取一包袋泡茶，放在杯里，用开水一冲就成了一杯又浓又香的红茶，非常方便。他们认为奶会掩盖茶叶本身的香气、滋味，因此酷爱清饮浓红茶。近年来，风味多样的调味茶，如红茶奶茶、草莓红茶、夏威夷果茶、薄荷绿茶、茉莉花茶等也越来越受欢迎。

Sri Lanka: The Nationwide Favor to Pure Black Tea

Sri Lanka was formerly known as "Ceylon". Ceylon Highland, together with China's Keemun and India's Assam and Darjeeling, are the world's four best-selling black teas. The tea business, since the import of tea seeds from China in 1824, has become a stable of Sri Lanka's agriculture over some 200 years. The country also comes top of the world's tea production and exports.

For Sri Lankans, tea is a part of life. Tea stalls spreading all over cities and rural areas provide a quick access to full-flavored black tea prepared simply with teabags and hot water. As milk is believed able to cover up tea's own fragrance and taste, they love pure, strong black tea. Recent years also saw the mounting prevalence of multi-flavored teas in Sri Lanka, such as milk tea, strawberry black tea, macadamia nut tea, mint green tea and jasmine tea.

英国茶俗，一切瞬间为茶而停

“当时钟敲响四下时，世上的一切瞬间为茶而停。”英国人饮茶历史已有 300 多年，是消费红茶最多的国家之一。1662 年，爱喝茶的葡萄牙公主凯瑟琳嫁给英国国王查理二世，将中国红茶和饮茶之风带入英国。

下午茶起源于 17 世纪，由于社交晚餐时间较晚，每天下午人们都感到饥饿和疲惫。一位女伯爵为了缓解饥饿，令人每天下午为她准备一壶红茶和一些点心，之后她邀请朋友一起喝茶，吃美味的三明治和传统的英式点心，谈天说地，享受下午茶时光。很快，下午茶在上流社会流传开来。18 世纪中期后，下午茶风尚逐渐平民化。红茶的冲泡方式、优雅的摆设和丰盛的茶点被视为下午茶的传统。

The UK: A Beverage for Which Everything Stops

"When the clock strikes four, everything stops for tea." For over three centuries have the British people been drinking tea, making the UK one of the largest black tea consumers in the world. In 1662, Catherine of Braganza married King Charles II of England, bringing her beloved Chinese black tea to her new homeland. There followed the drinking of tea as a vogue nationwide.

Afternoon tea appeared first in the 17th century. People in those years, in the habit of dining late, usually felt hungry and drained in the afternoon. To reduce hunger, a countess supposedly asked servants to prepare black tea and snacks every afternoon. She also invited friends to share the tea, sandwiches and vintage British refreshments. It seemed a really good time with tea, snack and chitchats. Soon was such "afternoon tea" in vogue among the British high society. By the mid-18th century, it turned popular with all segments of the British population. For quintessential British afternoon tea, dignified brewing methods, magnificent utensils and lavish finger foods are the hallmarks.

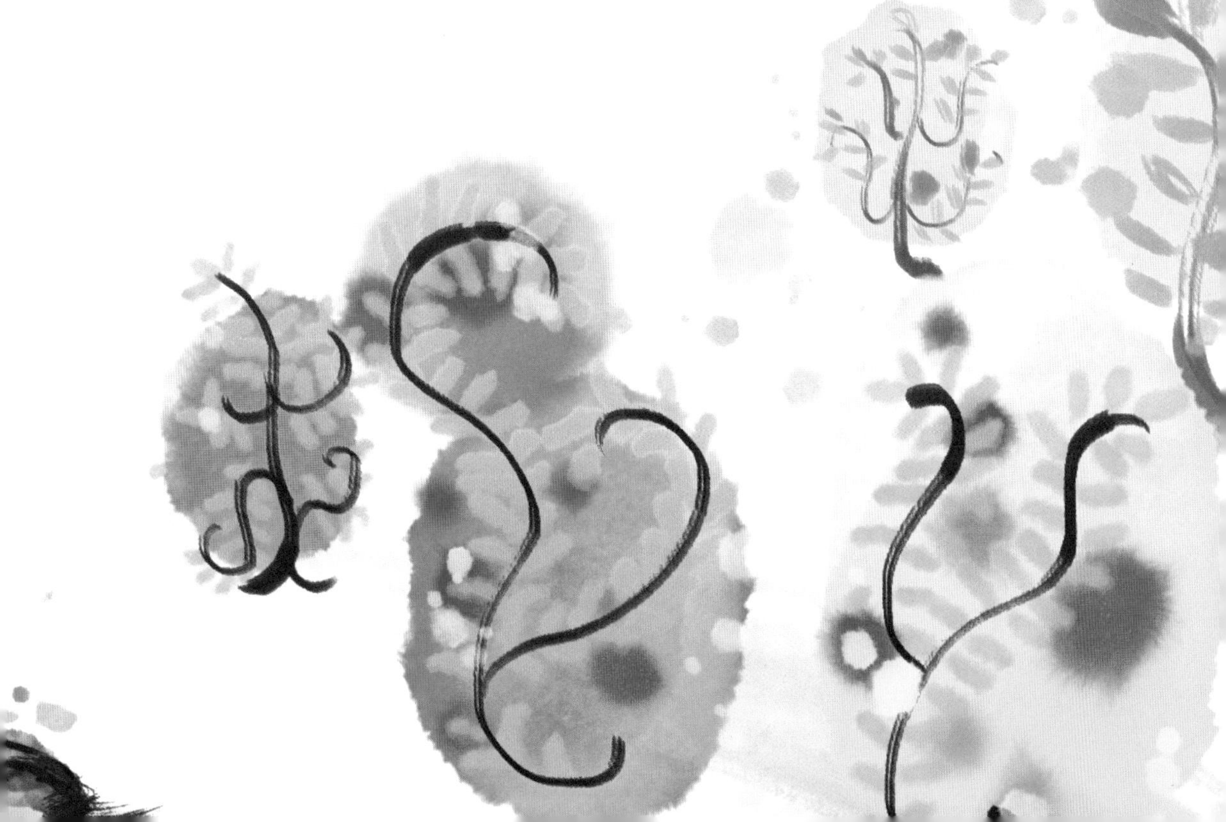

俄罗斯茶俗，茶炊与甜红茶

从 17 世纪后期开始，中国茶叶通过万里茶道源源不断地输入俄罗斯，俄罗斯饮茶之风逐渐由城市向各地普及，乡间茶会的内容就出现在俄罗斯著名诗人普希金的作品中。

俄罗斯传统的煮茶器具“茶炊”非常独特。茶炊的外形有球形、桶形、花瓶形等，被装饰得五彩缤纷。茶炊中间是一直桶，用于放木炭煮水，边上装有一个水龙头，水煮开后就从龙头放水泡茶。现代俄罗斯的城市家庭多使用茶壶泡茶，茶炊更具装饰作用。

俄罗斯人习惯饮用浓红茶，喜欢在茶里加方糖、柠檬片、蜂蜜、牛奶、果酱等。他们一日三餐离不开茶，尤其注重午餐，一顿丰盛的午餐过后还得上茶，饮茶时还要佐以饼干、奶渣饼、甜点和蛋糕等茶点心。

Russia: Samovar and Sweetened Black Tea

It was along the China-Russia Tea Road that tea from China were channeled into Russia from the late 17th century. Prevailing in big cities first, tea became later a popular drink over the rest of the country. Alexander Pushkin also put the countryside tea party in his literary works.

Samovar is the special tea-brewing urn traditionally used in Russia. It is usually flamboyantly designed, either shaped spherical or like a barrel orvase, containing a vertical charcoal pipe for heating water and a tap aside giving hot water to infuse. As teapots are much preferred in today's Russian cities, however, samovar is more like a finery.

Black tea has to be strong in Russia, and sugar cubes, lemon slices, honey, milk and jam are much favored supplements to enhance the tea flavor. Tea is never missing on Russians' dining tables. For them, lunch is decidedly important. An ample lunch must be followed by tea and some tea mates, like cookies, milk crumbs, desserts and cakes.

美国茶俗，更爱冰红茶

美国的饮茶历史有200多年，与其建国时间几乎等长，茶叶仅在夏威夷等地有少量种植，其余主要依靠进口，据2020年国际茶叶委员会统计，美国茶叶进口量名列世界第四，消费量名列世界第七。

美国人喜饮红茶、绿茶、乌龙茶等，红茶为首选，方便、快捷的速溶茶、袋泡茶备受欢迎。绝大多数美国人喜饮冰茶。冰茶制作简单，茶叶浸泡或煮沸后滤去茶渣，有的不加糖，有的加糖、柠檬或牛奶、果汁等调味，最后加入适量冰块，或放入冰箱冷却后饮用。

在美国的一些咖啡厅，除供应咖啡外，还供应红茶、绿茶、调味茶，并受到越来越多美国年轻人喜爱。现在，美国茶叶市场上中国草药茶和传统茶叶消费越来越多，特色茶逐渐走俏美国市场。

The US: Black Tea Prepared on the Rocks

Tea has remained popular in the United States for over two centuries, a time span almost the same as the American history. The country heavily relies on tea imports, with only the exception of few tea gardens in Hawaii and other places. Data from International Tea Committee (ITC) said that in 2020, the United States was the world's fourth largest tea importer and seventh largest tea consumer.

Black tea, green tea and oolong tea are the most favored sorts in the U.S, and black tea sells best. Americans prefer instant tea and teabags, and most of them love iced tea. It is simple to prepare it. By either steeping or cooking, we will have tea soup separated from the post-brewing grounds first. We may choose sugar, lemon, milk or juice to enrich the flavor, and then pour the mixture over ice cubes or get it chilled in the fridge. Some people, of course, prefer sugar-free iced tea.

Coffee aside, some cafés offer black tea, green tea and flavored tea. They have gained rising popularity among the young generations. In today's America, publicity starts to grow on exotic teas, like thc Chinese herbal tea and other heritage tea sorts.

摩洛哥茶俗，独爱甜绿茶

300 多年前，茶叶沿着丝绸之路，来到享有“北非花园”美誉的摩洛哥。摩洛哥地处非洲西北部地区，气候炎热而干燥，当地人喜食羊肉，爱好甜食，而饮茶能解油腻、助消化，因而茶叶是摩洛哥人不可或缺的生活必需品。

摩洛哥人主要饮用绿茶。沏茶时，他们先在壶里放入茶叶，冲少许沸水后立即倒掉，再冲入开水，放入很多白糖，加一大把鲜薄荷叶，泡几分钟后，才倒入杯子饮用。泡第二、第三次时，需酌量添加茶叶和糖。摩洛哥的茶具也很有特色，茶壶用银或铜制成，长嘴巴、大肚子，壶上有极富民族特色的图案。摩洛哥是世界上进口绿茶最多的国家，是个名副其实的爱茶国家。

Morocco: Green Tea Sweetened with Fresh Spearmint

Tea landed onto Morocco, a country usually prized as the "Garden of North Africa", along the Silk Road three centuries ago. Located at the northwest Africa, Morocco experiences scorching hot and dry weather all year round, and thus mutton and sweets are significant in people's diet. As tea helps digest greasy foods, it has become in this country a must-have in life.

Moroccans customarily drink green tea. They steep green tea first with some boiling water and immediately drain the pot, and prepare the essential brew with hot water and chunks of sugar and fresh mint leaves. The sweetened tea will be ready in a few minutes, and more tea and sugar must be in place for the second and third round. Moroccan tea sets also look special. The silver or copper teapot, with a gooseneck spout and a big belly, appealingly bears folk designs. As the world's largest green tea buyer, Morocco has a universal affection for tea.

新西兰茶俗，
红茶与多元茶饮融入

19 世纪中期，大量英国移民来到新西兰，他们带来了茶饮热潮（特别是红茶）和英式茶文化，影响了当地的饮食习惯。新西兰畜牧业发达，食物以奶、肉制品为主，茶能解油腻、助消化，饮茶成为新西兰人的生活必需。20 世纪 60 年代，新西兰人年均饮茶量仅次于英国。随着近 30 年来新移民的加入，亚洲各国的饮茶习俗也不断融入和丰富着新西兰的茶饮文化。

如今，新西兰人每日饮茶数次，年人均饮茶量居世界前列，无论是在企业还是在学校，均安排有茶歇。茶室、茶厅随处可见，人们在这里用餐，一般餐后都会供应奶茶或糖茶。一日三餐中，新西兰人特别重视晚餐，称其为“茶多”。

New Zealand: From Black Tea to More Varieties

The massive influx of British immigrants into New Zealand in the mid-19th century brought on the tea-drinking vogue (esp. black tea) and the British tea culture. It was even impactful to the locals' diet. People in New Zealand, thanks for the burgeoning animal husbandry, take dairy and meat most. Prized as grease-buster, tea has become a life necessity in this country. In the 1960s, New Zealand ranked next only to the UK in annual per capita tea consumption. New immigrants over the past three decades, particularly from Asia, have also helped grow the presence of their own tea customs in New Zealand.

People in New Zealand usually take "tea break" several times a day at all workplaces or even in schools. The country still ranks top when it comes to annual tea consumption per capita. Tea sweetened either with milk or sugar comes usually after meals, in particular the supper.

Chapter Five

—

The Festival of Tea
A Special Day for the Global Tea Community

第五部分

茶的节日

全球爱茶人的『国际茶日』

引文

茶叶是大自然赐予人类的不可或缺的瑰宝，是世界三大饮品之一，是连接人类文明的圣洁桥梁。2019 年第 74 届联合国大会将每年 5 月 21 日确定为“国际茶日”，倡导国际社会根据各国特色举办“国际茶日”庆祝活动，通过教育及宣传帮助公众更好地了解茶叶对农业发展和可持续生计等的重要性。

每年 5 月 21 日——“国际茶日”，让我们举起茶杯——茶和世界，共品共享!

Introduction

Tea, a heavenly gift inseparable from people's life, is the most-consumed beverage other than coffee and cocoa in the world. It also plays a blessed and pivotal role in the global cultural exchanges. Inaugurated by the 74th General Assembly in 2019, "International Tea Day" is a United Nations observance celebrated annually on May 21 worldwide with mobilized campaigns to enhance public awareness of tea's significant impact on global agriculture and sustainable livelihoods.

So may we have one voice on May 21— Tea and world, sharing and enjoying.

国际茶日，意义重大的爱茶人节日

继“世界粮食日”“世界土壤日”确立之后，农业领域又迎来一个重要的国际性节日——2019 年第 74 届联合国大会将每年 5 月 21 日确定为“国际茶日”，全球爱茶的人有了属于自己的节日！

全球 60 多个国家和地区产茶，茶叶年产量近 600 万吨，年贸易量 200 余万吨，饮茶人口超过 20 亿。茶叶是重要的经济作物，是部分最贫困国家主要的收入和出口创汇来源，对发展中国家的农村发展、减贫和粮食安全发挥重要作用。“国际茶日”的设立，有助于促进世界茶文化的交融互鉴和推动茶产业的协同发展，实现通过茶叶生产与加工减少极端贫困、应对饥饿、增强妇女权能、可持续利用陆地生态系统等重要目标。

A Festival for Global Ceberations

Following “World Food Day” and “World Soil Day”, another worldwide festival for agriculture came onto the scene. The global celebration of tea finally came true with the resolution of the 74th UN General Assembly in 2019 designating May 21 as “International Tea Day”.

Today’s global tea production from over 60 countries and regions has reached 6 million tons a year, and as for the volume of tea trade, the figure is 2 million tons. In fact, 2 billion people around the world drink tea. Tea is a significant cash crop, a main source of income and export revenues for some of the least developed countries. Tea can play a vital role in rural development, poverty reduction and food security in developing countries. The observance of International Tea Day can help facilitate the mutual understanding of worldwide tea cultures and the coordinated development of global tea industry, thus inspiring the contributions from tea production and processing to the reduction of extreme poverty, the fight against hunger, the empowerment of women and the sustainable use of terrestrial ecosystems.

国际茶日，茶叶故乡的深厚情怀

“芳茶冠六清，溢味播九区。”中国是茶的故乡，也是茶种植规模最大、茶产品种类最全、茶文化底蕴最深厚的国家，爱茶人大家族日益壮大。几千年来，茶不仅滋润着中华儿女的身心，还成为富裕一方百姓的“金叶子”。“国际茶日”是中国首次推动设立的农业领域国际性节日，体现了世界各国对中国茶文化的认可。

2020年5月21日首个“国际茶日”，中国国家主席习近平向“国际茶日”系列活动致信表示热烈祝贺：茶起源于中国，盛行于世界。联合国设立“国际茶日”，体现了国际社会对茶叶价值的认可与重视，对振兴茶产业、弘扬茶文化很有意义。作为茶叶生产和消费大国，中国愿同各方一道，推动全球茶产业持续健康发展，深化茶文化交融互鉴，让更多的人知茶、爱茶，共品茶香茶韵，共享美好生活。

A Cash Crop Speaking for China

"Tea spreading over the nine states puts all drinks in the shade," as an old Chinese poem said. China, the homeland of tea, is the world's largest tea producer. The country is so rich, and unmatchable, in tea cultures and varieties, and her population for a tea living is still vigorously growing. A full-flavored beverage to enlighten our souls aside, tea has become veritably a cash cow for local farming. International Tea Day, the first one China proposed for global observance in agricultural sector, asserts the identified Chinese thoughts of tea for worldwide good.

Chinese President Xi Jinping congratulated the launch of global celebrations for the first International Tea Day on May 21, 2020. In his letter, President Xi noted that"tea originated in China and later became popular around the world. The celebrations showed the international community's recognition of and emphasis on the value of tea, and they would be of great significance to revitalize tea industry and carry forward tea cultures. He stressed that China, a major tea producer and consumer, is willing to work with all sides for sustained and healthy development of the global tea industry, schedule more and wider tea culture exchanges, and find more people such an endearing beverage for a cheerful life."

国际茶日，在联合国切磋“世界茶道”

2020 年 5 月 21 日，中国常驻联合国代表团和联合国粮食及农业组织共同举办首个“国际茶日”线上庆祝活动，主题为“茶与可持续发展”，旨在凸显茶产业与减贫、消除饥饿、气候行动以及提升包容性等联合国重点行动领域之间的关系。联合国大会主席班迪、中国常驻联合国代表张军与来自俄罗斯、埃及、斯里兰卡、哈萨克斯坦、尼泊尔、肯尼亚、孟加拉国、匈牙利、黎巴嫩、荷兰等 20 多个国家的常驻联合国代表及近 200 位嘉宾“云聚一堂”，共同欢庆“世界茶日”。

“半壁山房待明月，一盏清茗酬知音。”中方现场展示了传统青瓷、紫砂茶具和各式茶叶，并通过穿插播放精美视频，同与会嘉宾“云分享”中国茶产业可持续发展的生动景象。茶文化、茶与减贫、茶产业应对新型冠状病毒感染疫情等专题视频受到各方一致称赞。嘉宾一致认为，茶叶促进减贫、茶产业抗击疫情的“中国故事”为其他国家特别是发展中国家提供了有益借鉴。俄罗斯、尼泊尔、肯尼亚也以视频形式展示了本国饮茶习俗和茶产业抗击疫情的情况。

A Celebration in the United Nations

Centered on "tea and sustainable development", the first International Tea Day was celebrated virtually by the Permanent Mission of China to the UN and the FAO on May 21, 2020. The connection between tea industry and the UN's prioritics, like poverty relief, the fight against hunger, climate actions and the promotion of inclusiveness, was put in the spotlight. The event featured the presence of the 74th UN General Assembly President Bande, China's UN-based Permanent Representative Zhang Jun and other UN representatives from over 20 countries, and nearly 200 guests from Russia, Egypt, Sri Lanka, Kazakhstan, Nepal, Kenya, Bangladesh, Hungary, Lebanon, the Netherlands, etc.

"A mild tea would be my best treat to bosom friends." The traditional Chinese celadon, purple-clay sets and the richness of tea sorts in short films showed to the world China's growing, sustainable tea economy. In particular, films on tea culture, poverty alleviation and China's fight against COVID-19 to resume tea production drew the attention of all parties. They all thought high of China's deeds and took a shine to the country's time-honored tea culture. In their view, "China's tea story" against poverty and the pandemic set a model for especially developing countries. Russia, Nepal, and Kenya also played films with reference to their tea customs and what to do to save their tea industry from the pandemic.

国际茶日，共品茶香、共叙茶谊

每逢“国际茶日”，全世界共品一杯茶，在美好的茶香茶韵中探讨一个共同的话题——不同文明和谐相生，“和而不同，美美与共”。

为了这份美好的愿望——

- 2021 年，中国率先在全球拉开“国际茶日”欢庆序幕。新春伊始就举办“大使品茶”“茶香农遗”等活动，邀请驻华大使品鉴“第一杯早春茶”，来自阿根廷、荷兰、斯洛文尼亚、阿联酋等 26 个国家和联合国

粮农组织、世界粮食计划署等国际组织的多位驻华使节和代表应邀出席。

• 2021 年 5 月 21 日，“国际茶日”中国主场活动与中国国际茶叶博览会联动举办，11 国驻华使节、200 多名中外嘉宾参与主场活动，联合国粮农组织总干事屈冬玉博士视频致辞，乌干达、伊朗、布基纳法索、爱尔兰、意大利等国嘉宾共聚一堂，全方位感受茶及茶文化的无穷魅力。

• 2022 年“国际茶日”，中国主场活动在工夫茶发源地广东潮州和“世界上早先的茶叶出产地”陕西安康举办，并邀请多国驻华使节和国际组织代表“云品香茗”，用一片茶叶诠释“和而不同，美美与共”的茶文化精神，传递不同文明和谐相生的人类命运共同体理念。

• 2022 年“国际茶日”，中国驻美使馆邀请美国政界、商界、文化机构等友人共品茶香、共叙茶谊，将以茶会友的待客之道从中国推向世界。

A Unity in Diversity for Worldwide Friendships

The co-existence of cultures, or "unity in diversity", has always been the absorbing topic of the global village observing the International Tea Day. China has been doing great for such worldwide friendships.

In early 2021, "the first spring tea" offered to ambassadors to China was the first session in global International Tea Day celebrations. Events at the beginning of the year, like "Tea and Ambassadors" and "Aroma in GIAHS Candidate Site", saw the presence of ambassadors from 26 countries, like Argentina, the Netherlands, Slovenia and the United Arab Emirates, and representatives from the FAO and WFP.

On May 21, 2021, China's main celebration was staged hand in hand with China International Tea Expo, which honorably had the presence of ambassadors from 11 countries, including Uganda, Iran, Burkina Faso, Ireland and Italy. Qu Dongyu, the FAO Director-General, also delivered a video speech to over 200 guests coming all the way to acquire the charms of tea and tea culture.

International Tea Day 2022 was noticeably celebrated in Chaozhou (Guangdong province), the home to Kungfu Tea, and Ankang (Shaanxi province), "one of the earliest tea-growing area in the world". Representatives from foreign embassies and international organizations on a video conference were invited to savor the tea fragrance and shared views about the "unity in diversity".

Meanwhile, the tea party hosted by the Chinese Embassy in the US, with the presence of lots of friends from political, business and cultural sectors, was another full-house blast to celebrate the International Tea Day.

国际茶日，共商全球茶产业发展

众人拾柴火焰高！“国际茶日”是凝聚茶人智慧，共商产业发展的平台。

2021 年 5 月 21 日，中国国际贸易促进会农业行业分会、中国茶产业联盟、意大利对外贸易委员会、斯里兰卡茶叶委员会等多家机构共同发起《“国际茶日”促进茶产业发展倡议》，呼吁全世界茶产业界——

- 以茶谊结君子之交；
- 以茶道求和而不同；
- 以茶技促绿色发展；
- 以茶义行共赢之道；
- 以茶园秀美丽乡村。

倡议得到多个国家和地区茶协、茶企的热烈响应。中国举办了茶业国际高峰论坛，全球茶产业精英共商茶文化、茶产业、茶科技统筹发展之路。

A Call for Development for All Stakeholders

"Many hands make light work." More of a tea community's annual conference, International Tea Day is a stage for consulting the future of the global tea industry.

An initiative for the promotion of tea industry was made public by CCPIT Agriculture, China Tea Industry Alliance, Italian Trade Commission, Sri Lankan Tea Board and other agencies on May 21, 2021 to call for worldwide friendship, mutual understanding, green development, win-win solutions and rural prosperity.

For tea fraternities around the world, the initiative appeared in good time. In China, a high-profile tea business forum was also staged to map out with the international tea community the future of the tea culture, industry and technology.

国际茶日，绘就茶农致富新画卷

“国际茶日”是促进茶叶消费、带动茶农增收、增加贸易机会的平台，多项拓展市场的积极尝试活跃了茶叶市场——

组织开展“全民饮茶月”“全民文化推广公益茶会”等大众活动，开设“国际茶日”线上讲堂，邀请茶专家、文化名人以茶知识倡导健康饮茶，以茶文化科普推动市场消费。

2020 年，为应对疫情给茶农带来的销售之困，组织茶叶主产区开展“国际茶日”直播带货活动，拓展消费渠道。5 月 21 日当天，直播间人数

突破千万，多款产品很快售罄，有效缓解了疫情对茶农生计的影响。

2021 年，“大使品茶”系列活动娓娓讲述浙江“莲城雾峰”的“早”、广东“英德红茶”的“香”、广西“六堡茶”的“乡愁”、福建“坦洋工夫”的“甜蜜”、四川“川红工夫”的“绝”……20 多种有特色、有故事的名茶引爆市场。

2022 年，为各国搭建推广平台，多国驻华使节、外国商协会代表和中国 12 个茶产区代表直播推介全球特色好物和中国好茶，见证茶与世界美食的精彩相遇，让世界共享中国市场发展机遇，推动茶叶这片“致富叶”成为改善全球茶农生计的“幸福叶”。

A Platform to Let Tea Farmers Earn More

As people also take International Tea Day as a chance for higher tea spending, earnings and trade volumes, China has done a lot more for spurring the tea sales.

Tea publicity events aside, like “Tea-Drinking Month” and “People's Cultural Promotion Tea Party”, lectures were virtually given by tea scientists for a spread of health knowledge, or by culture celebrities for a boost of spending power.

To combat the pandemic-caused sluggish sales in 2020, the state authority sponsored a live-stream fair for local tea farmers. On May 21 alone, viewers came in tens of millions and a number of stores ran out in seconds.The regional tea economy was thus rescued.

The “Tea and Ambassadors” series in 2021 unveiled to the public 20 or more much-sought teas, each with an unparalleled feature. “LianchengWufeng”, for instance, is a premium early-spring tea from Zhejiang; “Yingde Black Tea”, from Guangdong, is best known for its exceptionally pleasant aroma; Liubao Tea from Guangxi is an elixir to homesickness for overseas Chinese; Tanyang Kungfu Black Tea from Fujian stands for a lifetime “happiness”; and Chuanghong Kungfu Black Tea, named after its place of origin Sichuan, is considered a top-rated tea of China that can even rival the world-renowned Keemun.

Another attempt was made in 2022 with representatives from foreign embassies, trade unions and China’s 12 tea-producing areas for live-stream global specialty sales promotion. This was surprisingly a good talk between the world’s finest tea and food and a chance for the whole world to progress hand in hand with China, making the beverage a real cash cow for tea farmers around the globe.

国际茶日，跨越时空的文化名片

“国际茶日”是讲述各国茶故事、传承弘扬茶文化的交流平台，充分展现了茶融汇古今、贯通中西的包容性和延展性——

• 悠悠茶香、芳华少年，“大学生茶文化体验官”活动吸引百所高校大学生参观茶乡茶厂，学习茶艺茶道，感受茶文化的博大精深。

• 穿越时空、跨越国界，“红茶的世界传播史”纪录片带领人们沿着丝绸之路探寻红茶的前世今生。

• 采花成蜜、摘叶为茶，“5 · 21 国际茶日”和“5 · 20 世界蜜蜂日”同庆活动邀请多国嘉宾尝甘蜜、品香茗，见证茶蜜融合的美好情缘。

• 世界茶源，咖啡故乡，国际茶日云推介活动邀请咖啡发源地——埃塞俄比亚的客人共同开启一场茶与咖啡的起源对话。

A Stage for the Spread of Tea Legacies

International Tea Day takes the stage of cultural exchanges for spreading tea cultures all over the globe. Tea is more than a beverage globally accepted, it is a cultural phenomenon.

"Tea Culture Discovery", for example, was a pilgrimage of students from a hundred and more universities. They called on tea workshops, learned tea rituals and thus came to grab the kernel of the tea cultures.

"*The Global Travels of Black Tea*" was another attempt, a documentary in quest of the history of black tea along the Silk Road.

As May 20 is the World Bee Day, the co-celebration with International Tea Day featured the participation of people from around the world to savor the best honey and tea.

Like tea from China, coffee originated from Ethiopia. The two countries, on International Tea Day, held a grand web dialogue on the origins of tea and coffee.

国际茶日，世界茶人的深情告白

“国际茶日”这一天，全世界都沉浸在芬芳的茶香里，制茶、品茶、赛茶，充分展示着缤纷多彩的茶俗文化，尽情表达对茶的热爱。

联合国粮农组织举办“茶园到茶杯”等主题宣传活动；欧盟举办了内容丰富、形式多样的线上“国际茶日”讲解活动；英国推出“在 80 杯茶中环游世界”系列播客，讲述世界各地别具一格的茶风茶俗；斯里兰卡举办“茶艺大师风采大赛”，展示人与锡兰红茶的情缘；东南亚国家爱茶人各着民族服饰，展示其独特的茶饮习俗，庆祝和宣传“国际茶日”；日本以富有浓郁地域特色的“国际茶日暨静冈茶文化”主题宣讲活动庆祝“国际茶日”……此外，世界各地茶人茶客自发在网络上组织云品茶。

5 月 21 日，世界因茶而沸腾，人类因茶而和谐美好。

A Day for Global Tea Community Celebrations

International Tea Day is, of course, a global event. People around the world came in swarms to get grips on tea knowledge and customs that vary from country to country.

Celebrations came in different forms, like FAO's "from field to cup" thematic shows, online lectures and awareness sessions around EU members, the podcast series called "*Around the World in* 80 *Teas — the first few stops*" by the UK Tea & Infusions Association, the tea masters' competition in Sri Lanka for Ceylon black tea promotion, the costume and custom shows and celebrations in southeast Asia countries, the "Japanese Green Tea" cultural publicity celebrations by the prefecture of Shizuoka, and more tea drinking and networking activities of various sizes or levels around the world.

International Tea Day - a global celebration in the name of beauty, peace and harmony.